Experiments *in* Biochemistry: A Hands-on Approach

A Manual for the Undergraduate Laboratory

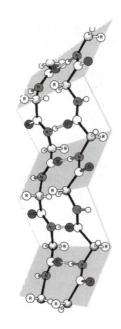

Shawn O. Farrell

and

Ryan T. Ranallo

Colorado State University

BROOKS/COLE

THOMSON LEARNING

Australia • Canada • Mexico • Singapore • Spain • United Kingdom • United States

For more information about our products,
contact us at:
Thomson Learning Academic Resource Center
1-800-423-0563

For permission to use material from this text,
contact us by:
Phone: 1-800-730-2214
Fax: 1-800-731-2215
Web: www.thomsonrights.com

Asia
Thomson Learning
60 Albert Complex, #15-01
Alpert Complex
Singapore 189969

Australia
Nelson Thomson Learning
102 Dodds Street
South Street
South Melbourne, Victoria 3205
Australia

Canada
Nelson Thomson Learning
1120 Birchmount Road
Toronto, Ontario M1K 5G4
Canada

Europe/Middle East/South Africa
Thomson Learning
Berkshire House
168-173 High Holborn
London WC1 V7AA
United Kingdom

Latin America
Thomson Learning
Seneca, 53
Colonia Polanco
11560 Mexico D.F.
Mexico

Spain
Paraninfo Thomson Learning
Calle/Magallanes, 25
28015 Madrid, Spain

Introduction to the Text

For Teachers and Students

Experiments in Biochemistry was written to complement a wide variety of biochemistry lecture formats. At Colorado State University, we have three levels of biochemistry lecture, with the choice based on career goals and major requirements. I have used Mary Campbell's *Biochemistry* for two different lecture classes and find that it contains the best level of detail and simplicity for those students. For that reason, *Experiments in Biochemistry* was written to complement Campbell's text as closely as possible; however, this manual is designed for use in independent lab courses as well as in conjunction with other biochemistry lecture texts.

Some lab classes are organized so that each experiment is independent. Often these classes meet only once a week and must provide the greatest overview possible in 9–15 weeks. Other classes, often for biochemistry majors, meet for longer times and more often. Such classes may wish to have experiments that flow more smoothly from one to another. For this reason, the experiments in *Experiments in Biochemistry* are divided into two types. For many sections there will be an experiment that stands alone followed by a similar experiment that can be part of an overall plan. These comprehensive experiments will usually have an experiment number followed by the letter "a." For most of the "a"-type experiments, it is assumed that the student will retain the sample collected during the previous experiment.

The model system for the comprehensive experiments is lactate dehydrogenase (LDH). An entire semester can be organized around the isolation, purification, and characterization of beef heart LDH, followed by the cloning and expression of recombinant vertebrate LDH in bacteria.

The methodology given is specific for the types of equipment we use in our lab, but any of these experiments could be optimized for similar but different devices.

Most of these experiments have been designed and improved by many of my students and teaching assistants and we are very happy with the results. All of the experiments are proven to work, and they can be completed in a normal lab period by a competent and prepared student.

We hope you enjoy doing these experiments as much as we have enjoyed designing them.

Acknowledgments

I would like to thank Mary K. Campbell, author of *Biochemistry,* and Saunders College Publishing for permission to use figures from the aforementioned text.

Special thanks go to Bio Rad Laboratories, Fisher Scientific, Milton-Roy, Calbiochem, Rainin Instruments, and Sigma Chemical Co. for permission to use figures from their instruction manuals in a previous version of this book.

The molecular biology portions of this manual would not have been possible without the generous assistance of Dr. Linda Holland from the University of California at San Diego.

Special thanks and congratulations to Cheryl Beseler for her work on the molecular biology portion of this manual and to Ryan Ranallo, who perfected the cloning experiment and helped proofread every chapter. Graduate students like Ryan and Cheryl are responsible for the success of these experiments in a teaching lab environment.

I would also like to thank the following reviewers who reviewed the manuscript and offered suggestions for revision:

Michael P. Garoutte, Missouri Southern State College

Robert N. Lindquist, San Francisco State University

Robert B. Sanders, University of Kansas

William M. Scovell, Bowling Green State University

Technical Help

An Instructor's Manual that includes instructor's guidelines to equipment and reagents is available to adopters of this book through your Brooks/Cole or Thomson Learning representative. Technical help concerning material in this manual is also available from the author. Feel free to contact Dr. Shawn Farrell at the following address:

141 MRB

Colorado State University

Ft. Collins, CO 80523

(970) 491-7916

sfarrell@lamar.colostate.edu

Table of Contents

Introduction to the Biochemistry Lab

Welcome to the biochemistry laboratory! Biochemistry is a fascinating subject that overlaps many other scientific fields. The techniques that you learn in a biochemistry lab will be applicable to all life sciences across a broad range of professional interests. Many courses in biology, microbiology, chemistry, botany, zoology, food science, and nutrition have laboratories that include large sections of biochemistry experiments and techniques. Whether your professional interest involves working in a lab as a researcher, going to graduate school, or entering a professional degree program such as medicine or veterinary medicine, skills and knowledge of biochemistry will be essential.

Objectives of the Biochemistry Laboratory

There are several objectives of a biochemistry laboratory. These include

1. Learning the physical skills and techniques of modern experimental biochemistry.

2. Learning how to think scientifically and independently, and do the requisite calculations.

3. Learning the theory behind the techniques and the biochemical pathways.

The physical skills are things like using a pipet correctly, properly using a pH meter, balancing centrifuge tubes, loading chromatography columns, setting up electrophoresis equipment, and so on. The techniques themselves refer to the actual types of experiments, each one of which has a specific goal and a proper time and place for its use. The skills and techniques are the most important part of this experience. Why is that? Because when you graduate, if you can list on your resume that you have done HPLC, SDS-PAGE, agarose gel electrophoresis of DNA, and so on, you will immediately impress potential employers. If you go into graduate school with a better than average understanding of these techniques, you will avoid needless repetition in your first-year graduate courses. That will put you a step ahead and get you into your real research faster. If you immediately go into a field such as medical technology, these techniques will be your livelihood.

Thinking scientifically and doing calculations is also an important part of laboratory science. You will never be just a pair of hands doing work—there will always be planning of future experiments and analyzing the data from the past ones. If you understand why things work the way they do, you will plan your experiments correctly. This will save you time and money in the long run. It would be a tragic shame if you did a brilliant, elegant experiment, and then got the wrong conclusions because of a simple math error. There are always calculations at the end of an experiment, and they are very, very important. Never say or think, "That didn't matter. It was just a math error. I understood the important stuff." In the medical field, which many of you may aspire to, simple math errors can be fatal. In the graduate laboratory, simple math errors can cost you several months' work and your boss several months of supplies.

The basic knowledge of biochemistry and the related pathways will be learned effectively in a lecture course, but hands-on experience in the lab is helpful to reinforce that knowledge. This is why all of the experiments revolve around some real biochemistry. Because you may not be taking this lab concurrently with a lecture course, I have tried to give all of the necessary background for each experiment, but it will be assumed that you have a basic biochemistry text, such as *Biochemistry* by Mary Campbell, or *Biochemistry* by Reginald Garrett and Charles Grisham. The theory behind the techniques is important so that the student knows how to apply the techniques in a new situation and plan experiments from the beginning.

Chapter Format of *Experiments in Biochemistry*

Each chapter is organized roughly according to the following topics:

Topics. This is what you should get out of reading each chapter and doing each experiment. If you have done the experiment and read the material carefully (perhaps more than once), and you still do not have a thorough understanding of the concepts, then one of us has not done our job well enough.

Introduction. This gives the background material that I would normally present in a lecture course studying the same topic. Sometimes the background may be more intensive than you need for the fairly simple technique that follows it, but it is better to be over-thorough in this case. Hopefully, between this manual and a basic biochemistry text, you will be able to complete all the experiments and figure out the answers to all the questions without spending time doing extensive research from other journals. The most important concepts will also be summarized in subsections called **Essential Information.** Problem-solving skills are demonstrated with the **Practice Sessions.** Sometimes there is a section called **Expanding the Topic,** which gives information that is not necessarily essential for completing the experiment, but goes into greater depth on some of the fine points.

Prelab. These are specific questions pertinent to the experiment you are about to do. I include them for a couple of reasons. First, if you have read the material, you should know how to answer these questions. If you do not, then you know you need to go back over it again. Second, if you do not know the answers to the questions, you are about to waste valuable lab time reading when you should be doing the experiment. That will put you behind, and you and your lab partner will probably be getting out of lab late. Third, your instructor may ask that you turn these in prior to lab as part of your lab grade. I know I always do.

Tip I.1

⮕ Half an hour in your instructor's office can save you hours of banging your head against your lab manual. If you cannot do your lab write-up in an hour or two, go get help from your TA or instructor. That is what they are paid for!

Experimental. This section gives you the materials and methods you will be using for the experiment in a step by step procedure. *Always, and I repeat, always read all of the procedures before beginning anything!*

Analysis of Results and Questions. This is where you report your results and do the calculations for your data.

Additional Problem Sets. Here you will find some more questions that your instructor may ask you to do, or you may just use them for practice.

Webconnections. Each chapter has a list of Web sites relevant to the topics covered. This list is available through the Thomson Learning Web site. You can access this site at http://www.brookscole.com by going to the *Experiments in Biochemistry* page located under Biochemistry on the Chemistry page.

References and Further Reading. These tell where much of the information in this manual came from. They are listed for professional etiquette reasons and not so you will run out and try to find them. You should be able to do all of the questions with just this manual and occasionally a basic biochemistry text.

Chapter 1

Biochemistry Boot Camp

Survival in the Biochemistry Lab

Topics

1.1 Lab Safety

1.2 Scientific Notation

1.3 Significant Figures

1.4 Statistics and Scientific Measurements

1.5 Units

1.6 Concentration of Solutions

1.7 Dilutions

1.8 Graphing

1.9 Pipets and Pipetmen®

Introduction

In this chapter, we discuss some fundamentals of laboratory science. This material is often overlooked by instructors who, correctly or incorrectly, assume that you have learned it well in other courses. It could be argued that the material presented here is the most important, however, due to the sheer magnitude of the repetition. You will constantly be doing calculations involving units, concentrations, and dilutions. You will be presenting your data using tables and graphs. You will do hundreds of pipettings during a semester. If this material is not mastered, you will pay a heavy price all semester. Read this chapter thoroughly, even if you think you already know it. You may just find an informational jewel you had not expected.

1.1 Lab Safety

A lab can be a dangerous place if you are not careful. This potential danger comes from several sources. First, the very nature of the chemicals and equipment used could be hazardous. We try to minimize use of hazardous materials, but some of the modern techniques in biochemistry require them. Hazardous chemicals are noted in the materials section of the experiments. Some equipment can be dangerous due to moving parts, heat, or potential electric shock. Only by using the equipment correctly, as instructed, will you be sure that you are safe.

Second, glassware used in a lab can always be considered dangerous because it breaks when dropped or mishandled. Flying glassware can come from anywhere, so you may not be the one who makes the mistake, but you may be the one who pays for it. Proper clothing and eye protection is the only sensible way to protect yourself against most common lab accidents.

Third, most accidents are caused by carelessness. A student who is mentally prepared to undertake the lab, has studied the material, and understands the procedures is much less likely to make a mistake that could injure someone. Simple mistakes like careless washing of glassware can be dangerous if water ends up on the floor and isn't cleaned up. Please note and observe the following lab procedures:

Things You Should Do:

1. Always use some form of protective eyewear. The most reliable are certified lab goggles that protect from the sides as well as the front. Contacts can be a problem in a lab because tears do not wash out things spilled in the eye when a contact is present. Eye wash stations are not efficient if you have contacts in. However, the American Chemical Society has recently removed its recommendation against contact lenses in the lab. Evidence shows that contacts are not dangerous if proper protection (goggles) is used.

2. Be aware of what chemicals you will be using. Wear gloves when using toxic chemicals. Remember that although you are not using a dangerous chemical, the student next to you on the bench may have spilled one.

3. Wear proper clothing in the lab. The lab is a good place for long sleeves, long pants, closed-toed shoes, and standard-lens glasses. It is a bad place for short sleeves, shorts, and sandals. Don't even think about going into a lab barefoot.

4. Familiarize yourself with the layout of the lab. Do you know where the fire extinguishers are? Where are the eye wash stations? Where is the first aid kit?

Things You Should Never Do:

1. Never eat, drink, or smoke in the lab. Although you may see more advanced scientists doing the first two in their labs, it is a bad idea in a teaching lab. There may be 200 students using that lab in a week, and the instructor cannot control what all of them are doing. If you need to drink often, then bring a water bottle, but leave it outside. Nobody will mind it if you step outside to eat or drink, but bringing any food or beverage into the lab, even if sealed, is a potential danger.

2. Never use mouth suction on glass pipets to draw up a solution. Even if you think the solution is a harmless buffer, the pipet may be contaminated with something hazardous. Use pipet pumps and bulbs to draw up solutions.

3. Never work alone. You should always have someone else in lab with you when you are working.

Additional Lab Courtesy

Although it is not strictly a safety issue, there is a lab etiquette you should follow as well. The following items will make your lab run more smoothly and lead to efficient transitions between lab sections:

1. Never stick your personal pipets into a community reagent bottle. If your pipet is dirty, you will have just contaminated the supply for the whole class. If you need 10 mL of a reagent, and there is a 1-L bottle in a community reagent area, take a small beaker over and pour in about 10 mL. That way, if your beaker is dirty, you will have contaminated only your own supply.

2. Never take a community reagent back to your own bench. One of the most frustrating things that can happen in a lab is to not be able to find something you need.

3. Don't use more of the chemical than you need. It seems that half of the students start taking chemicals from the community area before they have the slightest clue about how much they need. They seem to look at the size of the community reagent bottle and try to guess from there. Just because the reagent is out in a 1-L bottle doesn't mean you should take 100 mL of it. You might only need 2 mL. The 1-L bottle may be the total supply for all sections of the lab for the week.

4. Always clean up your lab area and any equipment and glassware you used. The next class may need to use the same stuff. The job is not over until the lab is clean and the equipment is ready for the next class.

1.2 Scientific Notation

You will probably see many numbers written many different ways during the course of your lab. The idea is to accurately and clearly communicate numerical information, and there is nothing necessarily wrong with any system of numbers that does that. However, there are some traditional ways of handling numbers that are used in science. One of these is called scientific notation.

Any number in **strict** scientific notation starts with one nonzero digit followed by a decimal point and some other numbers. That is followed by an exponent that tells to what power to raise it. Thus, the number 623 would be written 6.23×10^2. The number 0.0456 would be written as 4.56×10^{-2}.

When we are not using strict scientific notation, another common practice is to avoid starting a number with just a decimal point. Thus, rather than writing .453, we would write 0.453. When you label a test tube for storage, sometimes the ink fades, and it always seems as if the decimal point fades faster than the numbers. The beauty of using scientific notation is that if you used strict scientific notation, a fading decimal point wouldn't matter because you would always know that the decimal point belonged after the first number.

1.3   Significant Figures

It is important to realize that when you take a measurement or do a calculation, you can only rely on a certain amount of the information that you get from it. Analysis of significant figures is our way of figuring out how reliable the numbers are. Nowadays, many students use calculators or computers for everything, often with almost humorous results.

If you had a standard ruler with inches on one side and centimeters on the other, and you asked me to measure the height of your 250-mL beaker, what would you think if I told you it was 8.423587763 cm? You might say, "Thanks, Mr. Spock, but 8.4 cm would have been sufficient," but you would certainly see that I cannot measure something to that many decimal places when the smallest division on the ruler is 1 mm, or 0.1 cm. However, you might easily make a very similar mistake when you do a calculation. If you measured a cube to be 9.6 cm on a side and wanted to calculate the volume in cubic centimeters, you might multiply $9.6 \times 9.6 \times 9.6$ and report the volume as 884.736 cubic centimeters. Students do that all the time without recognizing that they were as wrong as the Mr. Spock example. If each side is measured to an accuracy of one decimal place, the answer cannot go to three decimal places.

Definition of Significant Figures

So, what is a significant figure? That has to do with the accuracy and precision of the instrument that you are using to make the measurement. If you have a balance, you weigh out a sample of histidine, and the balance reads 1.473 g, you have 4 significant figures. The last figure, the 3, is probably at the limit of the machine and is an estimate. If you weighed the same sample again, it might read 1.472 or 1.474. That would be like estimating between the closest marks on a ruler. Maybe a cruder balance would read 1.5 g, and you would have only two significant figures.

The number and position of zeros are often confusing to students with regard to significant figures. This is another reason to use scientific notation, because there is no ambiguity when you are using it. For example, how many significant figures are there in 0.0456? The answer is 3. Zeros before the first nonzero digit have nothing to do with the accuracy; rather they mark the place of the decimal point in scientific nomenclature. In scientific format this number would be 4.56×10^{-2}. How many significant figures are there in 0.045600? In this case, there are 5. The zeros following the six are significant and tell you about the accuracy of the measurement. In scientific notation this would be 4.5600×10^{-2}. In scientific notation, all zeros are significant. The real ambiguity comes when you write a number like 3200. How many significant figures are in 3200? You can't really tell. You could mean 3.2×10^2, or 3.20×10^2, or 3.200×10^2, which would be 2, 3, or 4 significant figures, respectively.

Significant Figures in Calculations

There are a couple of simple rules for using significant figures in calculations.

Multiplication and Division

When you multiply numbers, your answer will have the same number of significant figures as the number with the fewest. For example, if you wanted to calculate the volume of a cube by multiplying the length of the sides, it might look like this:

$$(3.4 \text{ cm}) \times (56.8 \text{ cm}) \times (2.435 \text{ cm}) = 470.2472 \quad \text{(on our calculator)}$$

How many significant figures can you claim in your answer? The answer is 2 because the first number you multiplied had only two significant figures. Thus the answer you should turn in is 4.7×10^2 cm^3. Division is done in exactly the same way.

Addition and Subtraction

Addition is a little different. When you add strings of numbers, you look at the number of decimal places to determine the accuracy of the measurement. The final answer cannot be more accurate than the least accurate number added. For example, if you are adding volumes to get a total, it might look like this:

$$(22.4 \text{ mL}) + (3.5 \times 10^2 \text{ mL}) + (0.543 \text{ mL}) = 372.943 \text{ mL}$$

How many significant figures can you claim? In this case, we can't really look at the significant figures in each term; rather, we look at decimal places. The figure with the fewest decimal places is 350. Therefore, our answer cannot go into tenths and hundredths. The true answer would be 373 mL. Note that the answer has 3 significant figures although one of the numbers added has only 2.

1.4 Statistics and Scientific Measurements

As a scientist you will deal with a great many numbers. We use statistics to help us get more meaning out of the numbers. An example any student can relate to would be test scores. If your friend tells you, "Hey, I got a 40 on the last exam," you would not immediately know whether you should be happy for your friend or not. How happy you would be might depend on several things, including what the total possible points were for the exam, what the average was on the exam, what the standard deviation was, and what you got on the exam.

You might also need to know whether you could really believe that number. Was the score added correctly? Was it a subjective score or objective one? To truly understand how that exam score relates to your friend's ability in that subject, you would also have to know whether it was likely that (s)he would get the same score again on a similar exam, or whether that score was overly high or low for some reason. When we make measurements during our experiments, we need to draw meaning from the numbers we see, and we often use statistics for this purpose.

If 200 students take the first exam in a class, we could list their scores as $x_1 = 85$, $x_2 = 64$, $x_3 = 98, \ldots, x_{200} = 12$. An individual number is usually designated x_i. What can we tell from these data? The students will want to know what their score means. If the professor is using a straight scale grading system in which 90–100% is an A, 80–89% is a B, and so on, then the first student (x_1) knows (s)he got a B. The third student knows (s)he got an A. The last student (x_{200}) knows that (s)he should consider dropping the class. If the professor is using a grading curve, then the students need to know some statistics to figure out how they are doing in the course.

There are two basic types of statistical measures. The first is a **measure of central tendency.** The one most often seen is the **average** or **arithmetic mean (x_{avg}).** The average is calculated by adding up all of the numbers and dividing by the total numbers you had.

$$\textbf{Arithmetic mean } (x_{avg}) = \frac{X_1 + X_2 + X_3 + \cdots + X_n}{n} = \frac{\Sigma X_i}{n}$$

Therefore, we would add up the scores for all of the students and divide by 200 to get the average on the first exam. If the average turned out to be 85, then the student with a 98 would feel pretty good. The student with an 85 would know (s)he was at the middle ground, and so on. On the other hand, if the average were only 40, then even the student scoring in the sixties would know that (s)he had done better than most. However, to calculate their current grades on a curve, the students would need more information.

A second type of statistical measure is a **measure of dispersion.** These usually measure the variations in the values compared to the mean. Often such a measure is necessary to truly understand the data presented. If I am the professor of the class taking the exam and I see that, of my 200 students, 180 of them scored an 85 on the exam, with a few being higher and a few lower, so that the average was still an 85, I will know that I have a very homogeneous class. Everybody did about the same with just a few exceptions. On the other hand, if the scores were scattered all over, with lots of students scoring below 40 and lots above 90, I will know that the class is heterogeneous. I would adapt my teaching strategy differently to the two classes. When grades are assigned on a curve, it is always the mean and some measure of dispersion that determines the grade.

There are many different measurements of dispersion, and the exact reason for using one or another is beyond the scope of this text. Some of the more common ones are presented next:

$$\textbf{Mean deviation} = \frac{\Sigma |x_i - x_{avg}|}{n}$$

where n is the number of samples measured and the $||$ means the absolute value of the difference between the individual value and the mean.

$$\text{Percent deviation} = \frac{\text{Mean deviation}}{\text{Mean}} \times 100$$

$$\text{Variance} = \sigma^2 = \frac{\Sigma(x_i - x_{avg})^2}{n - 1}$$

$$\text{Standard deviation} = \sigma = \sqrt{\sigma^2}$$

The mean and the standard deviation σ are the two most common statistical parameters. The students taking the exam would want to know their score in relation to the mean and the σ because if the class is graded on a strict bell curve, it usually takes a score of the mean $+ 1\sigma$ to give a grade of B and mean $+ 2\sigma$ to give a grade of A. Most students' calculators will determine mean and standard deviation for sets of data.

Errors in Experiments

You will often read about errors and error analysis in the data reported for an experiment. There are many sources of error in an experiment, and the term is often used as a catchall to explain numbers that are not perfect. It is important to realize that statistical error may not mean the same thing as the errors you are more familiar with. When you think of an error, you think of something like multiplying 6 times 9 and getting 42. When scientists think of error, they more often think of differences in members of a population. For example, if you measure the activity of the enzyme lactate dehydrogenase from the serum of 10 different people and express it as enzyme units per milliliter of blood serum, you will not get the same number twice. You might have a range of values from 0.1 units/mL to 26 units/mL. You could calculate the average and standard deviation and report your findings as

$$18.3 \pm 5.3$$

That would be error analysis, but the error does not imply that you made a mistake in your work. That is biological error, and simply reflects the individual variation in a population. You could have reported the values as mean $\pm$ variance or mean $\pm$ mean deviation. Either way would have given you different numbers.

If you attempt to pipet 1 mL with a Pipetman®, and you actually pipet 0.7 mL, there was error in the process. Was it caused by your inability to pipet correctly? Was the Pipetman® miscalibrated? In this case, there is a **true** value that is known, so it is easier to figure out the source of error. When sampling enzymes in the sera of biological organisms, it is harder to know the true value.

Accuracy *vs.* Precision

A measurement is **accurate** if it gives the true value. If you attempt to pipet 1 mL of water, which should weigh 1 g, and the balance reads 1 g after you dispense the solution, your pipetting was accurate. The arithmetic mean is the usual grounds for accuracy.

Measurements are **precise** if the same measurement can be made again and again. For example, if you try to dispense the 1 mL with the pipet ten times, and you dispense 0.7 mL ten times in a row, your pipetting was very precise, but it was inaccurate. This usually draws attention for the problem away from you and onto the equipment being used. It is the measurement of dispersion that is the usual criterion for precision.

Another parameter that can be measured in the case of pipetting is the **% error.** If you are trying to measure a known quantity, such as 1 mL of water and you expect it to weigh 1 g, the % error will give you a relative estimate of the error of your pipet or your pipetting technique.

$$\% \text{ error} = \frac{|x_{\text{avg}} - x_{\text{true}}|}{x_{\text{true}}} \times 100$$

So, in our pipetting example, if your average had been 0.7 g when it was supposed to be 1 g, the % error would be calculated as follows:

$$\% \text{ error} = \frac{|0.7 \text{ g} - 1.0 \text{ g}|}{1.0 \text{ g}} \times 100 = 30\%$$

1.5 Units

The international system of measurements is known as the SI (from *Systéme International d'Unités*), and is based on the MKS (meter-kilogram-second) system. The base units for SI are given in Table 1.1.

Many other units are derived from SI units, such as the joule ($m^2 \text{ kg/s}^2$), the unit of energy. Some non-SI units are also frequently used, such as degrees in Celsius (°C), pressure in atmospheres, and so on. The common unit of volume, liters, is not a SI unit, because volume would be cubic meters. A liter is actually a cubic decimeter.

Multiples of these units are most frequently used in biochemistry, and it is important that you know these like the proverbial back of your hand. Table 1.2 gives the most common multiples. With these multiples, we arrive at units such as millimeters (mm), micromoles (μmol), and so on.

Very, very few numbers in biochemistry exist without units. If you calculate the volume of something to be 12.34 cm^3, your answer will be wrong if you just report the number. It will be just as wrong as if you got the number wrong. It may seem like a small point to forget a unit, but think about the situation where the doctor tells the nurse to prepare a hypodermic needle with 10 of phenobarbital. What will the nurse prepare? Ten what? Ten mL Ten mg? If it is 10 mL, then 10 mL of what concentration? Is the drug absolutely pure or diluted? Without units, the nurse wouldn't know what to prepare.

Table 1.1 Basic SI Units		
Length	meter	m
Mass	kilogram	kg
Time	second	s
Temperature	Kelvin	K
Electric current	ampere	A
Amount	mole	mol
Radioactivity	Becquerel	Bq

Table 1.2 Prefixes for Multiple Units

Quantity	Prefix	Abbreviation
10^{12}	tera	T
10^9	giga	G
10^6	mega	M
10^3	kilo	k
10^{-1}	deci	d
10^{-2}	centi	c
10^{-3}	milli	m
10^{-6}	micro	μ
10^{-9}	nano	n
10^{-12}	pico	p
10^{-15}	femto	f
10^{-18}	atto	a

1.6   Concentration of Solutions

Every experiment you do will use at least one solution of a chemical. When chemicals are used as liquids, they can be either pure liquids or solutions. A pure liquid would be something like absolute ethanol. The chemical is a liquid, but every molecule in the bottle is ethanol with nothing mixed in. A solution contains a chemical dissolved in another liquid. When this is the case, you not only must know what the chemical of interest is, but you must know what it is dissolved in and how much of it is dissolved. The chemical in the smaller quantity is called the **solute.** The liquid it is dissolved in is called the **solvent.** When you calculate the amount of the solute that is dissolved in the solvent, you have determined the **concentration.** The concentration is very important to the biological or chemical function of a liquid. A good example would be if you made up a glass of Gatorade® from powder. If you follow the directions, you would put 1 scoop into an 8-oz glass or water bottle. If you instead put 3 scoops into the same 8 oz, you would still have Gatorade, but its concentration would be totally wrong. It would taste terrible and would not be emptied from your stomach properly. You might actually become dehydrated during your workout by using it.

Definition of Concentration

A concentration is always the ratio of an amount of a chemical divided by the total volume. It is important to be able to distinguish values that are concentrations from those that are not, which we will call amounts. Remember that amounts are additive, but concentrations are not. For example, if we put 1 g of salt in a beaker, that is an amount. If we bring the volume up to 1 L with water, then we have a 1 g/L solution, which is a concentration. If we have 1 mmol (millimole) of salt in a beaker, that is an amount. If we bring the volume up to 1 L with water, then we have a 1 mM (millimolar) solution, which is a concentration.

   If we have 1 g of salt in one beaker and 1 g of salt in another beaker, and we add them together, we will know that we have 2 g of salt. **Amounts are always additive.** If we have a solution that is 1 g/L and we add it to a solution that is 2 g/L, we do *not* get a solution that is 3 g/L. In fact, the solution would have a concentration in between 1 and 2 g/L based on the volumes that we added. **Concentrations are not additive.**

% Solutions

Solutions based on percent are the easiest to calculate, because they do not depend on a knowledge of the molecular weight. Your instructor can give you a tube with an unknown, white powder, and tell you to make up a 1% w/v solution, and you can do it without knowing anything about the white powder.

% w/v means **percent weight to volume** and has units of grams/100 mL. Therefore a 1% w/v solution has 1 g of solute in a total of 100 mL of solution.

% v/v means **percent volume to volume** and has units of mL/100 mL. Therefore a 1% v/v solution of ethanol has 1 mL of pure ethanol in 100 mL of total solution.

You can assume that unless told otherwise, the solvent is water for any solution.

> ### Tip 1.2
> Remember that when you calculate concentrations, the important thing is the final volume of the solution, not necessarily the amount you add. If you have 100 g of a solid chemical and add 1 L of water, the final volume will be more than 1 L. All calculations based on concentration assume you have correctly calculated the final volume after mixing the solute with the solvent.

Molar Solutions

The most common types of solutions are molar solutions. A 1 M solution means 1 mol of solute in a total volume of 1 L. A 1 mM solution has 1 mmol (10^{-3} mole) in a total of 1 L of solution. A 1 μM solution has 1 μmol (10^{-6} mole) of solute in a total of 1 L of solution.

Remember that moles are calculated by dividing grams by the formula weight in grams per mole. For example, if the formula weight of compound X is 39 g/mol, this means that one mole of compound X weighs 39 g. If we have 13 g of it, we calculate moles thus:

$$13 \text{ g} \div 39 \text{ g/mol} = 0.33 \text{ mol}$$

If we then take the 13 g and bring it to 0.5 L with water, our molar concentration will be

$$0.333 \text{ mol} \div 0.50 \text{ L} = 0.666 \text{ mol/L} = 0.67 \text{ M}$$

> ### Tip 1.3
> Remember that derivatives of molar solutions, such as mM, μM, nM, and so on still refer to an amount divided by liters of solution. Many students mistakenly think that a 1 mM solution means 1 mmol in 1 mL. What it actually means is 1 mmol in 1 L!

1.7 Dilutions

Dilutions seem to be one of the hardest concepts for most beginning biochemistry students. Even after chanting the mantra *"Dilutions are easy, dilutions are fun, dilutions make sense,"* students still struggle. Doing problems is the only way to really understand dilutions. A book such as Irwin Segel's *Biochemical Calculations* can also be a big help.

When doing a dilution, you always start with a concentration of something and add more solvent to it, thereby lowering the concentration. *Think about that!* If you do your calculations and determine that the concentration increased, you have made a mistake. Check your work each time by making sure that your concentration did, in fact, decrease with your dilution.

There are two simple ways of doing all dilution problems. Both methods are essentially the same if you understand math, but also have subtle differences that you may or may not appreciate.

$$C_1V_1 = C_2V_2$$

This method is good for those of you who are less comfortable with math. It is very reproducible and always works once you know how to define the variables. The disadvantage is that it is slower and you generate worthless intermediates if you have multiple dilutions. It is, however, a good starting point.

◄► Practice Session 1.1:

We have a sodium phosphate buffer at a concentration of 0.2 M (moles/L). We take 10 mL of the buffer and add it to 90 mL of water. What is the final concentration after the dilution?
First, to use the formula, you need 3 of the 4 variables.

C_1 = initial concentration, which in this case is 0.2 M

V_1 = initial volume, which is 10 mL

C_2 = final concentration, which is what we want to know

V_2 = final volume. *This is a potential danger point!* The final volume is 100 mL. Remember that the volume of solvent added (i.e. 90 mL) is only relevant in that it usually gives us the total volume. Sometimes 10 mL + 90 mL will not equal 100 mL. For the calculation you always use the total volume, so

$$C_1V_1 = C_2V_2$$
$$(0.2\ \text{M})(10\ \text{mL}) = C_2(100\ \text{mL})$$
$$C_2 = (0.2\ \text{M})(10\ \text{mL})/100\ \text{mL} = 0.02\ \text{M}$$

Dilution Factors

The second way is faster, and is better for multiple dilutions. I define a dilution factor to be the final volume divided by the initial volume. This is just a convention, but I use it consistently throughout this book. Therefore, in our previous example, the dilution factor is 100 mL/10 mL = 10, or a 10 to 1 dilution.

Now that we have the dilution factor, what do we do with it? Well, there are really only a couple of things we can do. We can multiply it by C_1, or we can divide it into C_1. One of the possibilities gives you an answer that is greater than what we started with, which we know can't be right. Therefore,

$$C_2 = C_1/D_f$$

$$C_2 = 0.2 \text{ M}/10 = 0.02 \text{ M}$$

I hope you can see that the two methods really are the same thing.

The reason that so many students get answers that indicate the concentration increased with dilution is because of nomenclature. Many books refer to a dilution as a 1 to 10, or 0.1. They get that by defining a dilution factor as initial volume over final volume. If you do that, there is nothing wrong with it, but you will have to then multiply your initial concentration by your dilution factor. Any way you do it that is consistent should work for you, but we will always be consistent and do it the way we described.

Multiple Dilutions

When you have multiple dilutions, it becomes more efficient to use the dilution factor method. If, for example, we made a dilution three times, the final concentration would be calculated thusly:

$$D_{f_1} = 100 \text{ mL}/10 \text{ mL} = 10$$

$$D_{f_2} = 50 \text{ mL}/2 \text{ mL} = 25$$

$$D_{f_3} = 90 \text{ mL}/30 \text{ mL} = 3$$

$$D_{f_{total}} = 10 \times 25 \times 3 = 750 \text{ to } 1$$

and if we had started with the same 0.2 M solution, the new concentration would be calculated as:

$$C_{final} = C_1/D_{f_{total}} = 0.2 \text{ M}/750 = 0.00027 \text{ M}$$

$$= 0.27 \text{ mM}$$

$$= 270 \text{ } \mu\text{M}$$

The final conversion from M to mM to μM will eventually be simple for you. Please refamiliarize yourself with the metric system so that you can do conversions effortlessly.

1.8   Graphing

Many of the experiments in this book will call for you to make graphs as you analyze your data, so we will spend some time reviewing graphing basics. The first part assumes you will be making the graph by hand instead of using a computer program, but it is important to understand these concepts before trying to make a computer do this for you.

Making Graphs by Hand

A graph is a plot of two quantities. One is something that you control and is called the **independent variable.** Usually it goes on the x-axis. The other is the quantity that changes as you change the independent variable. It is called the **dependent variable** and goes on the y-axis. For instance, if you put varying quantities of protein in a series of test tubes and then measure the absorbance, the variable you control is the quantity of protein, so it goes on the x-axis. The absorbance is what changes as you change the protein quantity, so it goes on the y-axis. By convention, "plot p vs. q" means that p is on the y-axis. **First and foremost, always use graph paper!** A graph scribbled on binder paper is not a graph.

Units and Exponents

Sometimes it is challenging to figure out what values to plot. Let's say we did a protein determination and recorded the following data:

Protein Concentration	Absorbance
0.001 mg	0.05
0.003 mg	0.16
0.005 mg	0.225
0.007 mg	0.33
0.008 mg	0.375

It would be particularly messy to plot those values in that form due to the number of zeros. The best thing to do is to change the units so easier numbers can be plotted. The easiest way is to plot μg instead of mg. That makes nice numbers like 1, 3, 5, 7, and 8 (see Figure 1.1). Another way is to use exponents. Most students of science make a mistake in using exponents because most graphs that you have seen have not been done with standard scientific nomenclature. Notice that the x-axis is labeled mg $\times 10^3$, *not* $\times 10^{-3}$. For scientific writing, the exponent indicates what you have already done to the number to put it on the graph with that scale, not what you expect the reader to do. In other words, the first value was 0.001 mg. To get rid of all of the zeros, we multiplied by 1000 to give a final value of 1. Therefore, on the axis we indicate that the value represents the number of milligrams of protein multiplied by 1000.

Drawing the Line

To make a graph, you must know something about the system and what you expect to see. When we do a protein assay, for example, we expect to get a straight line, at least to a certain limit of protein in the tube. We therefore try to draw the best straight line that we can through the points. You can do this by using a computer program to draw the line for you (see the next section), by using a calculator with a statistics function to tell you the slope of the line, or by visually putting down the line that goes through the most points. In Figure 1.1, it would be a mistake to just "connect the dots," thereby giving you a zigzag line.

When the relationship is expected to be linear, the best line is determined by **linear regression.** This is a mathematical process in which the line is placed to minimize the area

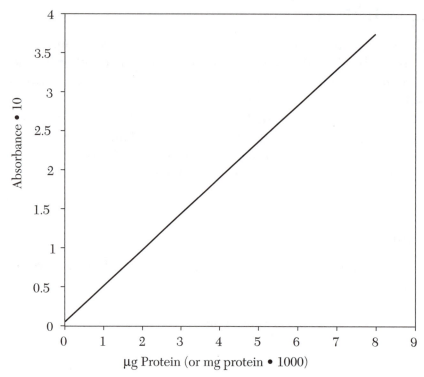

Figure 1.1 A proper linear graph

between the points and the line. This is also called the **least-squares** method. This is something that calculators and computers do easily, but your eyes do not.

Constant Scale

It is also important to remember that the scale must be constant. If you define 1 cm on the graph to be 1 μg as in Figure 1.1, then that scale must be maintained. A common mistake is to change the scale to fit your data instead of plotting your data on a given scale. This mistake is shown in Figure 1.2. Notice that this would give a very ugly line because the proper distance between the points was not maintained.

Use the Whole Graph

These first two figures are actually a bit small. It is generally best to use as much of the graph as possible. Also, the closer the line is to a 45° angle, the more accurate you will be when you interpolate an unknown from it.

Must Zero Be a Point on the Graph?

Most people naturally put the origin (0, 0) on the graph. Most of the time that will be correct, but there are times when it would be inappropriate. A common reason not to include zero is when you are graphing log molecular weight vs. mobility. This is done after doing an experiment with gel electrophoresis or gel filtration. For example, let's plot log MW vs. mobility for the following values:

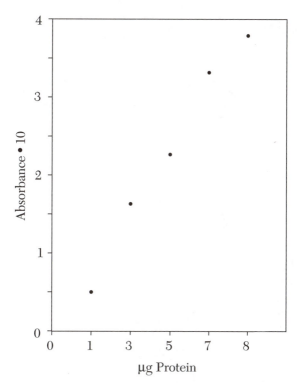

Figure 1.2 An improper graph

MW	Log MW	Mobility (cm)
15,000	4.18	6
40,000	4.60	2.9
66,000	4.82	1
90,000	4.95	0.1

If we start the graph at the origin, we will get a graph that looks like Figure 1.3. This graph is very compressed on the y-axis and does not use much of the graph paper. There would be huge error if we tried to interpolate within this graph. For this type of graph, it would have been much better to start the graph at 4 on the y-axis and go until 5, as in Figure 1.4.

Log Scales

You will make many graphs using a log scale such as in Figure 1.4. However, there is a much easier way to plot such data. You can buy some semilog paper, which will do the calculations for you. Log paper is divided into cycles. A cycle is for a set of data that are all within one power of 10. The data from the example given would fit onto one cycle because 15,000–90,000 are all within one power of 10. If we also wanted to plot the log of 200,000, then we would need a second cycle. With log paper, you just plot the numbers 15,000, 40,000, 66,000, 90,000, and 200,000 directly on the graph on the y-axis and the paper takes the log for you. **Do not take the log and try to plot 4.18, 4.60, . . . on log paper!** The graph in Figure 1.5 shows how these data would be plotted on two-cycle semilog paper.

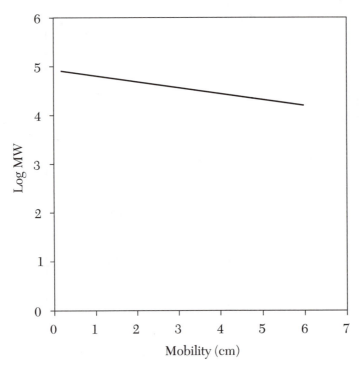

Figure 1.3 An unacceptable log-scale graph

Bad Data Points

The one thing that you can do manually better than a computer is decide when you have a bad data point. Let's say we want to plot the data for absorbance vs. μg protein shown in Table 1.3. If you plotted the points on a standard graph, you would most likely notice that 5 of the 6 points fall perfectly on a straight line. However, one point (30, 0.05) is way off. Most computer programs' default values would draw the graph shown in Figure 1.6 for these data. As you can see, the line is well below all of the data points except the bad one. Most of us would choose to redo the 30-μg tube rather than report data of this caliber. If that were not possible, we would probably draw the line through the 5 perfect points and ignore the bad point altogether.

Table 1.3 Absorbance vs. μg Protein

μg protein	Absorbance
0	0
10	0.10
20	0.20
30	0.05
40	0.40
50	0.50

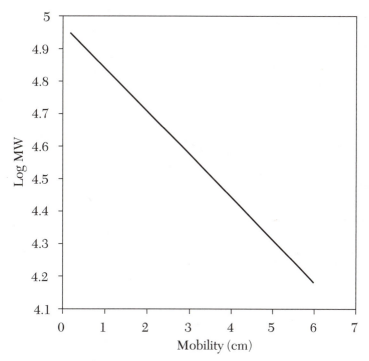

Figure 1.4 A proper log-scale graph

Graphing with Computers

Many of you probably have your own computer with a graphing program or spreadsheet with graphing capability. We live in the computer age, and more classes are becoming computer-driven. There are many graphing programs available. Sigmaplot™, Kaleida-graph™, Quattropro™, and Excel™ are just a few. We will use Excel as an example of how you can analyze data using spreadsheets and graphing programs.

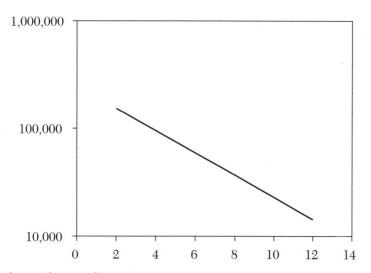

Figure 1.5 Two-cycle semilog graph

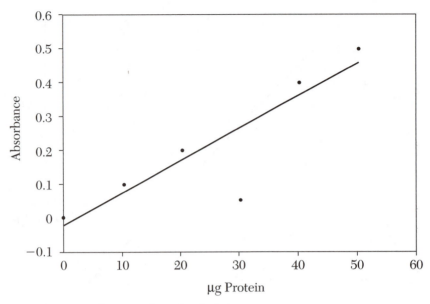

Figure 1.6 Linear regression for a graph with a bad data point

Loading the Spreadsheet

The first step is to set up the spreadsheet. With Excel you have a basic grid of columns, which are labeled A–Z, and rows, which are numbered one to several hundred. Each combination of a number and letter is called a **cell.** You start by putting data into the cells. Cells can hold data in the form of numbers or letters. For most applications that we are interested in, we put in numbers. If you wanted to plot absorbance vs. μg protein for the Bradford protein assay, you might have data that look like Table 1.4.

When you load this into Excel, you could put the column labels (μg Protein and Absorbance) into the spreadsheet, but that is not necessary when you are planning to use the data to make a graph. The program will talk you through creation of the graph, including labeling the axes. It is best to put the values that will be on the x-axis into the left-hand column, as it simplifies the graphing process. Figure 1.7 shows what the spreadsheet would look like.

Table 1.4	Absorbance versus μg Protein for the Bradford Assay	

μg Protein	Absorbance
0	0
10	0.10
20	0.21
30	0.29
40	0.40
50	0.52

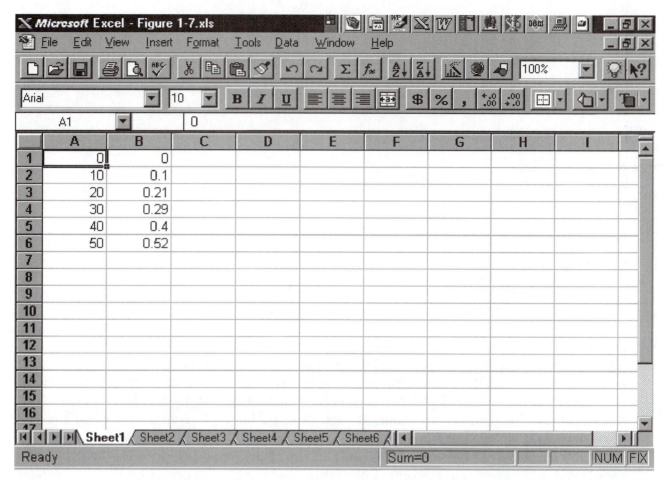

Figure 1.7   Excel spreadsheet for linear graph demonstration

Creating the Graph

Once you have loaded the data into the spreadsheet, you are ready to create the graph. With Excel, it is a user-friendly process involving dialog boxes. First you have to select the data that will be put into the graph, so you click on the A1 cell and drag down and over until all of your data are highlighted. Then you click on the chart wizard button. The first dialog box that pops up gives you a choice of chart types, such as bar graph, pie graph, scatter plot, and so on. We will choose to do the XY scatter plot, which is normally what you will want to do. *Here is a potential danger in using a computer to make a graph!* The default value for many computer programs would have straight lines connecting the points (the dreaded "connect the dot" graph). This is usually wrong in biochemistry. What you want is a best-fit line of some type. In our case, it would be a linear regression line. At this point in the process, we choose the XY scatter that does not attempt to put any line over the points. The line will be put in later. With whatever program you are using, you must learn how to make the program give you the type of graph you want. Don't rely on the default values. With Excel, we would click on "Next" to continue the dialog. The next box that pops up (box 2 of 4) asks about chart source data. If you have correctly selected the data from the columns you wanted,

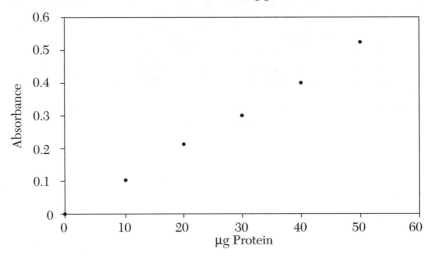

Figure 1.8 Absorbance versus μg protein for Excel graphing demonstration

you can go on to the next box by clicking on "Next." This brings us to box 3 of 4 in the dialog process. This is the chart-formatting step where you can decide how you want the graph to look. There is a "title" tab you can select and then add in the graph title and the title of both axes. There is a "gridline" tab you can select that allows you to decide how many gridlines, if any, will appear on the graph. At each step, a minigraph picture shows you how the graph currently looks, so you can continue to make changes until it looks the way you want. Any time you change your mind, you can click on "Back" to go back to the last box. The fourth dialog box gives you the option of putting the graph as an object on the spreadsheet itself or making it a separate page. Select the latter because it gives a better formatted page that you can then print easily. After so doing, the graph we have been working on would look like Figure 1.8.

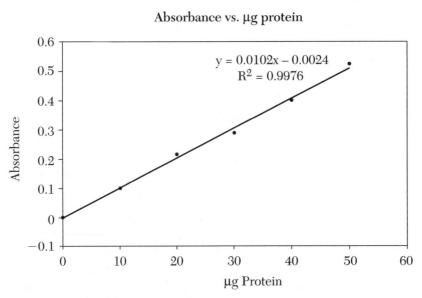

Figure 1.9 Customized graph of data presented in Figure 1–7

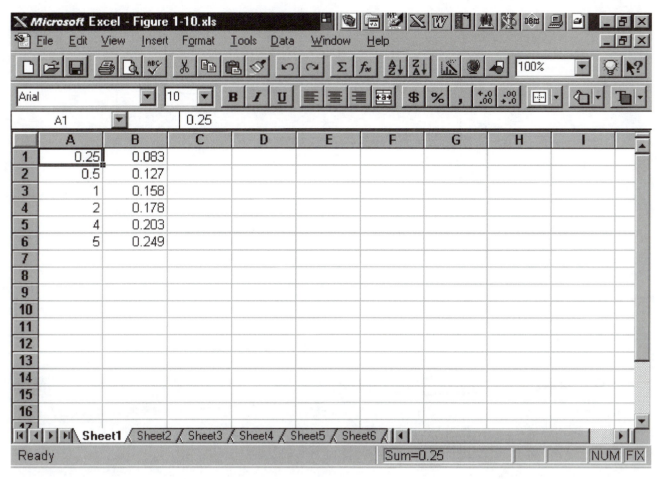

Figure 1.10 Excel spreadsheet for hyperbolic graph

To avoid creating the default "connect the dots" graph, we selected a graph that had no line. We still have to create the line. This is the one place in Excel that is not user friendly. You actually have to know something about the program, or you have to use the help menu. In this case, we create the line using the pulldown menus on the top bar. We choose "chart" and then the "add trendline" choice from the pulldown menu. The box that pops up gives you a picture of possible types of lines, including regression and polynomial. If we choose regression, we get the type of line we want for this type of linear data. An "options" tab also lets us choose to put the regression coefficient and the mathematical equation of the line on the graph. We would end up with the graph shown in Figure 1.9.

Nonlinear Graphs

What do you do if the graph does not depict a linear relationship? Making such a graph would start out the same way, but then you would use the chart wizard differently to customize your graph. A common example of this would be analyzing an enzyme kinetics experiment, such as in Chapter 8. When enzyme velocity is plotted vs. substrate concentration, a hyperbolic relationship is seen. Figure 1.10 shows typical enzyme kinetic data as they would appear in an Excel spreadsheet. The left column represents the concentrations of substrate in millimoles/liter, and the right column is the enzyme velocity in micromoles per minute.

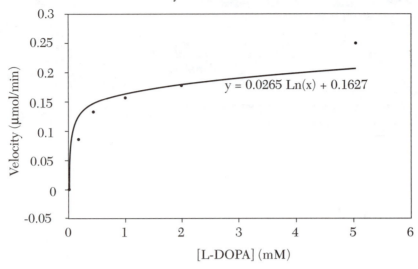

Velocity vs. Substrate Concentration

$$y = 0.0265\ \text{Ln}(x) + 0.1627$$

Velocity (μmol/min)

[L-DOPA] (mM)

Figure 1.11 Incorrect logarithmic graph of enzyme kinetic data

	A	B	C	D	E	F	G	H	I
1	0.25	0.083	4	12.04819					
2	0.5	0.127	2	7.874016					
3	1	0.158	1	6.329114					
4	2	0.178	0.5	5.617978					
5	4	0.203	0.25	4.926108					
6	5	0.249	0.2	4.016064					

Figure 1.12 Excel spreadsheet for linear transformation

Lineweaver–Burk Reciprocal Plot

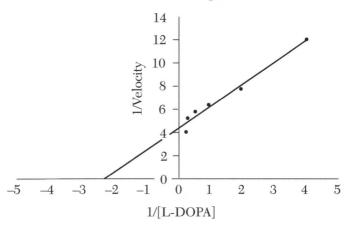

1/[L-DOPA]

Figure 1.13 Reciprocal graph of enzyme kinetic data

Once the original graph, without a line, is created, you would add a trendline as before. For hyperbolic graphs, there is no correct menu choice, unfortunately. The best you could do would be to select "logarithmic" when you get to the trendline. This would give you the graph shown in Figure 1.11.

This graph looks reasonable, except that it is a logarithmic function. A logarithmic graph is curved and similar to a hyperbolic one, but the equations do not match. This graph would give incorrect values if you attempted to interpolate kinetic constants from them. The good news is that you can use your spreadsheet to manipulate the data. If you take the reciprocals of the data and plot those, you will be plotting a Lineweaver–Burk graph. Figure 1.12 shows the Excel spreadsheet with the reciprocal transformation. Figure 1.13 shows the completed Lineweaver–Burk graph. By using trendline options, it was possible to extend the line until it intercepted the *x*-axis. This is necessary to measure the kinetic constants from this graph.

1.9 Pipets and Pipetmen®

Much of your success in the biochem lab revolves around your ability to choose and use various devices for measuring and dispensing solutions. These range from simple glass tubes, called pipets, to highly advanced and expensive units collectively referred to as Pipetmen, although this is really a name given to a particular company's product, similar to saying Kleenex to mean any kind of facial tissue.

As you learn to use these liquid transfer devices, you should always keep in mind that there are two things you are striving for—accuracy and precision. **Accuracy** is the relation between the volume you dispense and the volume you wanted to dispense. If you have a pipet, you are trying to dispense 100 μL, and you actually dispense 70 μL, your volume is very inaccurate. **Precision** has to do with reproducibility. If you are supposed to dispense 100 μL into each of 5 tubes, and what you actually dispense is 70.0 μL, 69.9 μL, 70.2 μL, 70.1 μL, and 69.8 μL, then your volumes are very inaccurate but very precise. Many of the

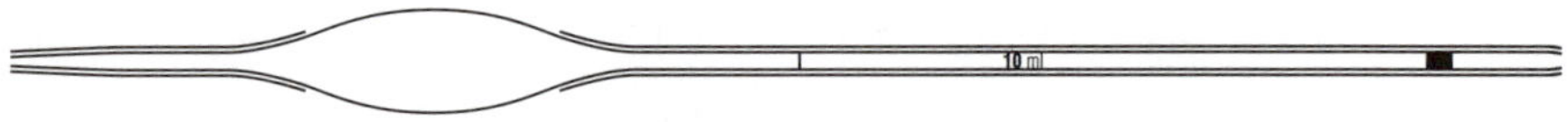

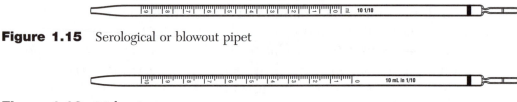

Figure 1.14 Volumetric pipet

mechanized Pipetmen-type dispensers are famous for their precision, but their accuracy depends greatly on your technique and how often they are calibrated.

Volumetric Pipets

A volumetric pipet is calibrated to deliver one set volume. An example is shown in Figure 1.14. The tolerance for error is sometimes written on the pipet. For example, a 10-mL pipet may say ±0.2%, and that gives the expected accuracy of the pipet.

Measuring Pipets

These pipets are graduated to indicate varying volumes of liquid, so you can use them to deliver many different volumes up to the maximum volume of the pipet. There are two basic types. **Serological** pipets, often called **blowout** pipets, are graduated all the way to the tip (see Figure 1.15). If you have a 10-mL serological pipet and you want to deliver 10 mL, you bring the volume all the way up to the zero line and then expel all of the liquid. **Mohr** pipets are not graduated to the tip, so to dispense the same 10 mL, you would let the meniscus run down from the zero line and then stop it at the 10 line (see Figure 1.16).

Because your accuracy depends on how well you can get the meniscus to start and stop on the lines you want, a pipet is most accurate when used closest to its full capacity. A simple calculation will verify this. Suppose you are going to dispense 10-mL using a Mohr pipet. The pipet has graduations of 0.1 mL. If your hand and eye coordination is such that you routinely have an error of ±0.1 mL, you might accidentally stop the meniscus on the 9.9 line. That would be an error of 0.1 mL/9.9 mL = 0.01 or 1%. Now, if you tried to use the same 10-mL pipet to dispense 1 mL, you would draw the solution up to the zero line and attempt to stop it at the 1-mL mark. If you had the same inaccuracy in your technique, you would actually stop it at the 0.9-mL mark. The error would be 0.1 mL/0.9 mL = 0.11 or 11%. That is why you should always choose the pipet that you can fill the most. It makes sense to use that 10-mL pipet to pipet anything from 5.1 to 10 mL, but if you want to pipet only 4 mL, then you should choose the common 5-mL pipet. If you wanted to pipet 2 mL, then the 5-mL pipet would not be so good, because there is also a common 2-mL pipet, and so on. Make sure you know what pipets are available and use them near to capacity.

Figure 1.15 Serological or blowout pipet

Figure 1.16 Mohr pipet

To use glass pipets correctly, follow these procedures:

1. Using a pipet pump or rubber bulb, draw solution into the pipet up past the mark you want to start at.

2. Depending on the device you used to suck up the solution, you may either leave the device attached or remove it and use your index finger to control the flow.

3. Wipe off the tip with a tissue.

4. Touch the tip of the pipet to the sides of the vessel you took the solution out of and let it run down until the meniscus is at the line you want to start at.

5. Transfer the pipet to the vessel you want to put the solution into and touch it to the sides of the vessel.

6. Let the solution run down the sides until the meniscus is at the line you want to stop at.

Glass Syringes

One of the most accurate devices is the glass syringe, the most common of which is the Hamilton® syringe (see Figure 1.17). These have a tight-fitting metal plunger in a glass tube and often are used for accurate dispensing of volumes as small as 1 μL.

To use a Hamilton syringe correctly, follow these procedures:

1. Draw solution into the syringe and expel it several times to wet the inside of the glass and the needle.

2. Slowly draw solution up until the plunger is at the line you want.

3. Watch carefully for bubbles forming. If you get bubbles between the solution and the plunger, expel the solution quickly to shoot the bubble out.

4. Repeat until you have no bubbles in the syringe.

5. Wash the syringe out repeatedly with water. Never leave salts or organics in the syringe, or the plunger will become permanently locked.

6. Be careful not to bend the plungers. They bend easily, and once bent are useless.

Pipetmen

The generic term for these is air-displacement piston pipets, so you can see why we say Pipetmen. The three most common types in use these days are the Pipetman by Rainin Instruments, the Eppendorf® pipettor by Brinkman Instruments, and the Integrapette® by

Figure 1.17   Hamilton syringe
(Courtesy of Hamilton Co.)

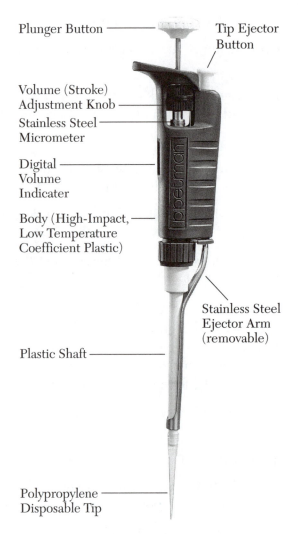

Plunger Button

Tip Ejector
Button

Volume (Stroke)
Adjustment Knob

Stainless Steel
Micrometer

Digital
Volume
Indicater

Body (High-Impact,
Low Temperature
Coefficient Plastic)

Stainless Steel
Ejector Arm
(removable)

Plastic Shaft

Polypropylene
Disposable Tip

Figure 1.18 The Rainin Pipetman
(Courtesy of Rainin Instruments, Inc.)

Integrated Instrument Services. We will use the Pipetman as an example throughout this section (see Figure 1.18).

Pipetmen come in sizes from the P-2, with a maximum volume of 2 μL to the P-5000, with a maximum volume of 5 mL. The most common sizes are the P-20, P-200, and P-1000. The instrument uses disposable plastic tips in various sizes depending on the size of the Pipetman. There is a color-coding system between the pipet and the tips used. For example, the P-1000 has a blue top to indicate it should use blue pipet tips, which are the size that goes up to 1 mL. The P-20 and P-200 both have yellow tops and use yellow tips, which go up to 200 μL. Before you start, make sure you have the proper size Pipetman and the proper tips to use.

All of the Pipetmen are infinitely variable from a lower limit up to the maximum. Never try to dial the Pipetman past its maximum, because this will irreversibly damage the unit. As you set the volume on the pipetman, you must keep in mind what the maximum is. The volume dial has 3 digits, and what those three digits mean depends on the total volume possible for the pipet. For example, if the dial reads 125, that would be 125 μL if it is a P-1000,

Table 1.5 Volume Ranges of Rainin Pipetmen

| Pipetman Model | Range (μL) | | Smallest Increment (μL) |
	Adjustable	Recommended	
P-2	0 to 2	0.1 to 2	0.002
P-10	0 to 10	0.5 to 10	0.02
P-20	0 to 20	2 to 20	0.02
P-100	0 to 100	10 to 100	0.2
P-200	0 to 200	50 to 200	0.2
P-1000	0 to 1000	100 to 1000	2.0
P-5000	0 to 5000	500 to 5000	2.0

or 12.5 μL if it is a P-20. Table 1.5 shows the recommended range for the Rainin Pipetmen as well as a description of the dial readings for each one. The other brands of pipettors are similar.

Accuracy of Pipetmen

It is very easy to establish the accuracy of a pipettor. All you have to do is pipet a known volume of water onto a balance and weigh it. If you have a digital top loader balance in the lab, this takes but a few seconds and could save you hours of scratching your head later. In a teaching lab it is especially important to verify that the instrument you are about to use is accurate. If you have a P-1000, 1000 μL or 1 mL should weigh 1 g. If you have a P-200, put on 200 μL and it should weigh 0.2 g, and so on.

How to Use a Pipetman

1. Make sure you have the correct size Pipetman and tips.

2. Place the tip on the pipettor and tighten with your fingers, being careful not to touch the very tip.

3. Hold the pipettor vertically over the solution you want to suck up. Push the plunger down to the first stop. Be careful to not drive past the first stop.

4. Put the tip into the solution to a depth of a few millimeters.

5. Slowly let the plunger come up as solution is drawn into the tip. After the plunger is all the way up, hold the tip in the solution for a couple of seconds to make sure solution is no longer moving.

Tip 1.4

Prewetting the pipet tip can help if you are pipetting liquids with a low surface tension, such as organics. They tend to drip out of the pipet tip the first time you load them. Draw up the organic, then blow it out. Draw up some more, and this time it should hold without dripping.

6. Look at the tip to make sure you drew up the correct amount of solution.

7. Move the Pipetman to the vessel you want to dispense into. Touch the tip to the side of the vessel.

8. Slowly depress the plunger to the first stop, then past the first stop to the second stop. This will blow the solution out of the tip. Any solution remaining should not be blown out further.

9. Use the tip ejector to dispose of the tip.

ESSENTIAL INFORMATION

Nearly all scientific calculations involve units. An answer without units is a wrong answer. We use units based on those of the MKS (meter-kilogram-second) system. These units often have prefixes to tell you the power of ten based on the standard unit. Learn how to convert from one prefix to another without needing to write long conversion factors. You will need to do this hundreds of times during a biochemistry course. Memorize the most-used prefixes, such as kilo, milli, micro, and nano.

Your results will depend largely on how well you pipet. Learn how to use all of the glass pipets and automatic pipettors available. You must know when to use a particular liquid transfer device and when not to. Practice drawing up standard volumes into the Pipetmen. Then, during an experiment, you will know if you have pipetted the correct amount.

Study the section on graphing carefully. You cannot make a graph if you do not know what the relationship is between the x- and y-coordinates. Always use graph paper or a computer graphing program. Never scribble a graph on regular binder paper. You should rarely connect the dots on a graph. You should almost always draw a best-fit curve. Be wary of the default values on your computer program. Make the computer work for you, not the other way around.

Experiment 1

Use of Pipettors

In this experiment, you will learn how to use Pipetmen of various sizes and measure their accuracy, precision, and calibration.

Prelab Questions

1. What is the usable range of a P-1000 Rainin Pipetman?

2. What is the difference between accuracy and precision?

3. What should 100 μL of water weigh?

4. What should 1000 μL of water weigh?

Objectives

1. To become familiar and adept at using pipettors

2. To be able to check the calibration of the pipettors

Experimental Procedures

Materials

> 1000-μL pipettors
>
> 100- to 200-μL pipettors
>
> Yellow and blue pipet tips
>
> Deionized water
>
> Balances

Methods

Part A: Precision of P-100 Pipettors

1. Acquire one of the P-100– or P-200–type pipettors with the correct size tips. Make sure the color matches.

2. Set the pipettor to 100 μL.

3. Place weighing paper or a weighing boat on the balance and tare the weight to zero.

4. Draw up the designated volume of deionized water and dispense it onto the weighing paper or weighing boat. Record the weight of the water.

5. Repeat the procedure twice more.

6. Draw up 10, 20, 50, and 75 μL just to see what they look like in the tip.

Part B: Precision of P-1000 Pipettors

1. Acquire one of the P-1000–type pipettors and the correct size tips. Make sure the color matches.

2. Set the pipettor to 1000 μL.

3. Place weighing paper or a weighing boat on the balance and tare the weight to zero.

4. Draw up the designated volume of deionized water and dispense it onto the weighing paper or weighing boat. Record the weight of the water.

5. Repeat the procedure twice more.

6. Set the pipettor to 100 μL and check the weight of the liquid three times.

7. Draw up 200, 500, and 750 μL of water just to see what they look like in the tip.

8. Check the analysis of results questions, make sure you have all the data you need, and put the pipettors away.

Name _____ Section _____

Lab partner(s) _____ Date _____

Analysis of Results
Experiment 1: Use of Pipettors

Part A: Precision of P-100 Pipettors

1. Record the weight you measured for the three trials of 100 μL:

    Weight 1 (x_1) _____

    Weight 2 (x_2) _____

    Weight 3 (x_3) _____

2. Average the three weights.

    Average of three trials: _____

3. Calculate the % error between the average of the three trials and the true value:

    $$\% \text{ Error} = \frac{|\text{Avg weight} - 0.100 \text{ g}|}{0.1 \text{ g}} \times 100 = \underline{\hspace{2cm}}$$

4. Calculate the mean deviation for the three trials:

    $$\text{Mean deviation} = \frac{\Sigma |x_i - x_{\text{avg}}|}{3} = \underline{\hspace{2cm}}$$

Part B: Precision of P-1000 Pipettors

1. Record the weight you measured for the three trials of 1000 μL:

    Weight 1 (x_1) _____

Weight 2 (x_2) _____

Weight 3 (x_3) _____

2. Average the three weights.

   Average of three trials: _____

3. Calculate the % error between the average of the three trials and the true value:

   $$\% \text{ Error} = \frac{|\text{Avg weight} - 1.00 \text{ g}|}{1.00 \text{ g}} \times 100 = \underline{\hspace{2cm}}$$

4. Calculate the mean deviation for the three trials:

   $$\text{Mean deviation} = \frac{\Sigma |x_i - x_{\text{avg}}|}{3} = \underline{\hspace{2cm}}$$

5. Record the weight you measured for the three trials of 100 μL using the P-1000:

   Weight 1 (x_1) _____

   Weight 2 (x_2) _____

   Weight 3 (x_3) _____

6. Average the three weights.

   Average of three trials: _____

7. Calculate the % error between the average of the three trials and the true value.

   $$\% \text{ Error} = \frac{|\text{Avg weight} - 0.100 \text{ g}|}{0.1 \text{ g}} \times 100 = \underline{\hspace{2cm}}$$

8. Calculate the mean deviation for the three trials:

   $$\text{Mean deviation} = \frac{\Sigma |x_i - x_{\text{avg}}|}{3} = \underline{\hspace{2cm}}$$

9. Which of the two pipettors you used was the more accurate? Explain.

10. Which of the two pipettors you used was the more precise? Explain.

11. What are the take-home messages from this exercise? Give three specific things that you learned from this lab.

Additional Problem Set

1. How many grams of solid NaOH are required to prepare 200 mL of a 0.05 M solution?

2. What would be the concentration from problem 1 expressed in % w/v?

3. How many mL of 5 M NaCl are required to prepare 1500 mL of 0.002 M NaCl?

4. What would be the concentration of the diluted solution from problem 3 expressed in mM, μM, and nM?

5. A solution contains 15 g of $CaCl_2$ in a total volume of 190 mL. Express the concentration in terms of g/L, % w/v, M, and mM.

6. Given stock solutions of glucose (1 M), asparagine (100 mM), and NaH_2PO_4 (50 mM), how much of each solution would you need to prepare 500 mL of a reagent that contains 0.05 M glucose, 10 mM asparagine, and 2 mM NaH_2PO_4?

7. Calculate the number of millimoles in 500 mg of each of the following amino acids: alanine (MW = 89), leucine (131), tryptophan (204), cysteine (121), and glutamic acid (147).

8. What molarity of HCl is needed so that 5 mL diluted to 300 mL will yield 0.2 M?

9. How much 0.2 M HCl can be made from 5.0 mL of 12.0 M HCl solution?

10. What weight of glucose is required to prepare 2 L of a 5% w/v solution?

11. How many mL of an 8.56% solution can be prepared from 42.8 g of sucrose?

12. How many mL of $CHCl_3$ are needed to prepare a 2.5% v/v solution in 500 mL of methanol?

13. If a 250-mL solution of ethanol in water is prepared with 4 mL of absolute ethanol, what is the concentration of ethanol in % v/v?

Webconnections

For a list of Web sites related to the material covered in this chapter, go to **Webconnections** at the *Experiments in Biochemistry* site on the Saunders College Publishing Web page. You can access this page at http://www.saunderscollege.com. **Webconnections** are under *Experiments in Biochemistry* in the Biochemistry portion of the Chemistry page.

References and Further Reading

R. F. Boyer, *Modern Experimental Biochemistry*, Addison-Wesley, 1993.

W. S. Cleveland, *The Elements of Graphing Data*, Wadsworth Publishers, 1985.

R. C. Jack, *Basic Biochemical Laboratory Procedures and Computing*, Oxford University Press, 1995.

Rainin Instrument Root Co., Inc. *Instructions for Pipetman®*, 1991.

J. F. Robyt and B. J. White, *Biochemical Techniques*, Waveland Press, 1990.

I. H. Segel, *Biochemical Calculations*, Wiley Interscience, 1976. Still the best source of practice problems in basic biochemistry.

R. G. D. Steel and J. H. Torrie, *Principles and Procedures of Statistics: A Biometrical Approach*, 2nd ed., McGraw-Hill, 1980.

Chapter 2

Acids, Bases, and Buffers

Topics

Introduction

All biochemical reactions occur under conditions of strict control over the concentration of hydrogen ion. Biological life cannot withstand large changes in hydrogen ion concentration, which we measure as the pH. When chemicals exist that keep the pH from changing, we say the system is buffered. Whether in your body or in a test tube, the reactions that will be important to you will be buffered. The proper choice and preparation of a buffer is paramount to your success in a biochemistry lab.

Before we can truly understand buffers, however, we must understand the more basic concepts of strong vs. weak acids and pH. Throughout this text, we will use the Bronsted definition of *acid* as a substance that can donate a hydrogen ion and a *base* as a substance that can accept a hydrogen ion.

2.1 Strong Acids and Bases

An acid is a compound that has a hydrogen ion that it can give up to the solution. Common strong acids with which you are familiar are hydrochloric acid (HCl), sulfuric acid (H_2SO_4), and nitric acid (HNO_3). When strong acids are dissolved in water, they dissociate completely into their ions:

$$HCl \ (aq) \longrightarrow H^+ \ (aq) + Cl^- \ (aq)$$

A more formal way of writing this equation would indicate that the hydrogen ion is not really hanging out loose. There are no "naked" protons, as we say; rather, the hydrogen ion is attached to a water molecule to give a hydronium ion:

$$HCl + H_2O \longrightarrow H_3O^+(aq) + Cl^-(aq)$$

We'll use the abbreviated formulas for simplicity's sake.

Strong bases, such as NaOH, KOH, and $Ca(OH)_2$ likewise dissociate completely into ions.

$$NaOH \longrightarrow Na^+ + OH^-$$

The strength of an acid is determined by how much of the hydrogen ion dissociates when the acid is put into water. This can be determined from the K_a.

$$K_a = \frac{[H^+][A^-]}{[HA]}$$

The complete dissociation is indicated by the unidirectional arrows. If we calculated the K_a for the HCl, it would be much greater than 1. If we did the same thing for the NaOH (using the corresponding K_b), it would also be very large.

Because strong acids and strong bases break down completely into their ions, it is relatively easy to calculate the pH. As long as you know how many moles of starting compound you have, you will know how many ions you get. To calculate the pH of a solution of a strong acid or strong base, we use the following procedure.

▭▷ Practice Session 2.1

What is the pH of 0.01 M HCl?

Because HCl dissociates completely, the concentration of the H^+ is also 0.01 M.

$$pH = -\log[H^+] = -\log 0.01 = 2$$

Answer

$$pH = 2$$

▭▷ Practice Session 2.2

What is the pH of 0.01 M NaOH?

NaOH is a strong base, so it dissociates completely into Na^+ and OH^- ions. Therefore, the concentration of OH^- is also 0.01 M.

This is also easy, because we know that the product of the concentration of hydrogen ion and the concentration of hydroxide ion is always equal to 10^{-14}. This is called the water equation.

$$[H^+][OH^-] = 10^{-14} \, M^2$$

$$[H^+] = 10^{-14} \, M^2/0.01 \, M = 10^{-12}M$$

$$pH = -\log 10^{-12} = 12$$

Answer

$$pH = 12$$

2.2 Weak Acids and Bases

Determining the pH of solutions of weak acids or bases is a little trickier. Because they do not dissociate completely, determining the $[H^+]$ is more difficult, and an equilibrium expression with K_a must be used. The K_a tells us how strong the acid is, so it can be used to calculate how much H^+ dissociates from the acid. We would soon tire of writing equilibrium constants such as 10^{-8}, 4.3×10^{-6} and so on, so we have simplified matters by using pK_as. Basically the "p" of anything is $-\log$ of that quantity, just as the pH is $-\log[H^+]$. Therefore:

$$pK_a = -\log K_a$$

So if $K_a = 10^{-12}$, $pK_a = 12$. Because all reactions can be written in two ways, it is important to remember that when you use K_as or pK_as, you must write the reaction as an acid dissociation.

How you figure the pH will depend on which chemical species you have in solution. For a simple weak acid system, there are three possibilities: **weak acid only, weak base only, or buffer.** We will only consider the first two here.

Weak Acid Only Situations

If we have 0.002 mol of the weak acid, acetic acid, and we bring the volume up to 100 mL with pure water, we will have a solution of HAc at 0.02 M. Some of that HAc will break down via the following equation:

$$HAc \rightleftharpoons H^+ + Ac^-$$

How much will break down? We don't know because it is a weak acid and doesn't break down completely. The pK_a is a measure of acid strength and indirectly will tell us how much breaks down. Shortcut formulas are available to do these calculations:

$$pH = \frac{pK_a - \log[HA]}{2}$$

$$pH = \frac{4.76 - \log(0.02)}{2}$$

$$pH = 3.23$$

Remember, you use this equation when the problem gives you the amount or concentration of a weak acid and you have no way of figuring out the amount of H^+ or A^-. This short cut formula works for most reasonable concentrations of acids that you would see in biochemistry. The shortcut formulas are valid for concentrations up to about 0.1 M and for pK_a values as low as about 3. Once the acid becomes much stronger than that or the total concentration is higher, the shortcut formulas start to break down. See Section 2.11 for more info on these.

Weak Base Only Situations

What if we started out with sodium acetate, NaAc. This is the weak base formed from acetic acid. If we start with a weak base, some of it will react with water via the following equation:

$$NaAc + H_2O \rightleftharpoons Na^+ + HAc + OH^-$$

How much will react this way? Again, if this had been a strong base, the calculation would be simple, but because it is a weak base, we are not sure to what extent the reaction occurs. The pK_a indirectly tells us this. There is another equation to use for this situation:

$$pH = \frac{pK_a + 14 + \log[A^-]}{2}$$

$$pH = \frac{4.76 + 14 + \log(0.02)}{2}$$

$$pH = 8.53$$

2.3 Polyprotic Acids

The use of polyprotic acids, those with more than one hydrogen that can dissociate, creates one other complication. Each dissociation of a hydrogen has its own dissociation constant. If you look at the table of pK_as in this chapter, you will see several weak acids with more than one pK_a listed. This means they are polyprotic acids. Not to worry! The beauty of the system is that you normally only have to worry about one of them at a time. For example, if you start out with H_3PO_4, you have a weak acid only situation, and you will be using the 2.12 pKa. If you start out with Na_3PO_4, you have a weak base only situation, and you will be using the 12.32 pK_a.

Unfortunately, there is one other possibility, and that is called an **intermediate of a polyprotic acid.** If you put 0.01 mol of NaH_2PO_4 into water, what type of solution do you have? The active species is $H_2PO_4^-$, but what is it? It could be the acid in the following equation:

$$H_2PO_4^- \rightleftharpoons H^+ + HPO_4^{2-}$$

or it could be the base in the following equation:

$$H_2PO_4^- + H_2O \rightleftharpoons H_3PO_4 + OH^-$$

As it turns out, when you have an intermediate of a polyprotic acid, the pH is controlled solely by the two pK_a values and is independent of the concentration of the solution. The pH would be calculated by the following equation:

$$pH = \frac{pK_a1 + pK_a2}{2}$$

2.4 Buffers

What happens if we add both HAc and NaAc to the same solution? This is the definition of a buffer because it has both the weak acid and the weak base in the same solution. Some of the HAc would tend to break down via the equation

$$HAc \rightleftharpoons H^+ + Ac^-$$

and some of the NaAC would tend to break down via the other equation:

$$NaAc + H_2O \rightleftharpoons Na^+ + HAc + OH^-$$

This would be very confusing were it not for the fact that with weak acid systems these two competing reactions tend to cancel each other out and these reactions do not happen to any great extent because they are so weak.

Tip 2.1

A buffer is a solution that contains both the weak acid form and the weak base form. Your success with the write-up for this chapter will depend on how well you recognize whether a solution is a weak acid, a weak base, or a buffer.

The Henderson–Hasselbalch equation can be used to figure out the pH when you have a buffer:

$$pH = pK_a + \log(A^-/HA)$$

So, if we added 0.001 mol of HAc to 0.002 mol of NaAc and brought the volume up to 100 mL with pure water, the pH would be calculated as:

$$pH = 4.76 + \log(0.002/0.001)$$

$$pH = 5.06$$

You may have noticed that although we used three different equations for three different situations, we used the same pK_a each time. This is because the pK_a is really a number that is a constant for a reaction rather than a particular molecule. In each case, we were dealing with acetic acid and acetate. The pK_a of 4.76 is the pKa of the reaction

$$HAc \rightleftharpoons H^+ + Ac^-$$

Why a Buffer Is a Buffer

Buffers resist pH changes because they use up excess hydrogen ion or hydroxide ion. If we have a solution with both a weak acid and its salt, and we add some H^+, then the following reaction occurs:

$$A^- + H^+ \longrightarrow HA$$

Conversely, if we add OH^-, the following occurs:

$$HA + OH^- \longrightarrow A^- + H_2O$$

◕▭▷ Practice Session 2.3

Consider the following two systems:

System A: 0.1 L of pure water at pH 7

System B: 0.1 L of 0.1 M phosphate buffer at pH 7.2

If we now add 10 mL of 0.001 M HCl to each, what will be the pH in each solution?

System A.   10 mL = 0.01 L of 0.001 M HCl

$$0.001 \text{ mol/L} \times 0.01 \text{ L} = 1 \times 10^{-5} \text{ mol } H^+$$

The final volume is now 0.1 L + 0.01 L = 0.11 L, so the $[H^+] = 1 \times 10^{-5}$ mol/0.11 L = 9.09×10^{-5} M H^+.

$$\textbf{pH} = \mathbf{-log\ 9.09 \times 10^{-5} = 4.04}$$

System B.   At pH 7.2, which is the second pK_a for phosphoric acid, HA = A^-.

$$0.1 \text{ M phosphate buffer} \times 0.1 \text{ L} = 0.01 \text{ mol phosphate}$$

Because HA = A^- and the total is 0.01 mol, there must be 0.005 mol of each form of phosphate, or 0.005 mol $H_2PO_4^-$ and 0.005 mol HPO_4^{2-}.
   Initially

$$pH = pK_a + \log([HPO_4^{2-}]/[H_2PO_4^-])$$

$$7.2 \quad 7.2 + \log (0.005/0.005)$$

Now we add 1×10^{-5} mol H^+. For every mole of H^+ added, the following reaction occurs:

$$H^+ + HPO_4^{2-} \longrightarrow H_2PO_4^-$$

Therefore, 1×10^{-5} mol of $H_2PO_4^-$ will be created and 1×10^{-5} moles HPO_4^{2-} will be lost. Now the Henderson–Hasselbalch equation looks like this:

$$pH = 7.2 + \log (4.99 \times 10^{-3}/5.01 \times 10^{-3})$$

$$\textbf{pH = 7.198}$$

As you can see, the buffer changed from pH 7.2 to 7.198 with the added acid while the unbuffered system plummeted to pH 4.04. Had you been trying to run a reaction in

A buffer is a solution of a weak acid and its conjugate base. It resists pH change when there are reasonable amounts of both forms present. A buffer is best when used close to its pK_a. Buffers are made by taking a weak acid and titrating with strong base until the pH is correct or taking a weak base and adding strong acid until the pH is correct. A buffer can protect against pH changes from added hydrogen ion or hydroxide ion as long as there is sufficient basic form and acid form, respectively. As soon as you run out of one of the forms, you no longer have a buffer. To be a good buffer, the pH of the solution must be within one pH unit of the pK_a. The pK_a gives an idea of the strength of the acid and tells us what the pH will be when there are equal amounts of acid form and basic form present.

water, your system's pH would have plummeted and your experiment would have been ruined.

2.5 Good's Buffers

The original buffers used in the lab were made from simple weak acids and bases, such as acetic acid, phosphoric acid, and citric acid. It was eventually discovered that many of these buffers had limitations. They often changed their pH too much if the solution was diluted or if the temperature was changed. They often permeated cells in solution, thereby changing the chemistry of the interior of the cell. A scientist named N. E. Good developed a series of buffers that are zwitterions, molecules with both a positive and negative charge. These do not readily permeate cell membranes. They also are more resistant to concentration and temperature changes.

Most of the common synthetic buffers used today have strange names that you will quickly forget and even stranger structures. Table 2.1 gives a few examples.

The important thing to remember is that you don't really need to know the structure in order to use a buffer correctly. The important considerations are the pK_a of the buffer and the concentration you want to have. The Henderson–Hasselbalch equation works just fine whether or not you know the structure of the compound in question.

➥ Practice Session 2.4

What is the pH of a solution if you mix 100 mL of 0.2 M HEPES in the acid form with 200 mL of 0.2 M HEPES in the basic form?

You have both the acid and base forms, so you have a buffer and need to use the Henderson–Hasselbalch equation. First you must find the pK_a, which we know is 7.55. Then you must calculate the ratio of the base to acid. The formula calls for the concentration, but in this situation, the ratio of the concentrations will be the same as the ratio of the moles, which will be the same as the ratio of the volumes because both solutions had the same starting concentration of 0.2. Thus, we can see that the ratio of base to acid is 2/1 because we added twice the volume of base.

$$pH = pK_a + \log([A^-]/[HA]) = 7.55 + \log(2) = \mathbf{7.85}$$

Table 2.1 Acid and Base Forms of Some Useful Biochemical Buffers

Acid Form		Base Form	pK_a
TRIS—H$^+$ (protonated form) $(HOCH_2)_3CNH_3^+$	N—tris[hydroxymethyl]aminomethane (TRIS) $\rightleftharpoons$	TRIS (free amine) $(HOCH_2)_3CNH_2$	8.3
$^-$TES—H$^+$ (zwitterionic form) $(HOCH_2)_3C\overset{+}{N}H_2CH_2CH_2SO_3^-$	N—tris[hydroxymethyl]methyl-2-aminoethane sulfonate (TES) $\rightleftharpoons$	$^-$TES (anionic form) $(HOCH_2)_3CNHCH_2CH_2SO_3^-$	7.55
$^-$HEPES—H$^+$ (zwitterionic form)	N—2—hydroxyethylpiperazine-N'-2-ethane sulfonate (HEPES) $\rightleftharpoons$	$^-$HEPES (anionic form)	7.55
$HOCH_2CH_2\overset{+}{N}$ $NCH_2CH_2SO_3^-$ (ring, H)		$HOCH_2CH_2N$ $NCH_2CH_2SO_3^-$ (ring)	
$^-$MOPS—H$^+$ (zwitterionic form)	3—[N—morpholino]propane-sulfonic acid (MOPS) $\rightleftharpoons$	$^-$MOPS (anionic form)	7.2
O $^+NCH_2CH_2CH_2SO_3^-$ (ring, H)		O $NCH_2CH_2CH_2SO_3^-$ (ring)	
$^{2-}$PIPES—H$^+$ (protonated dianion)	Piperazine—N,N'-bis[2-ethanesulfonic acid] (PIPES) $\rightleftharpoons$	$^{2-}$PIPES (dianion)	6.8
$^-O_3SCH_2CH_2N$ $^+NCH_2CH_2SO_3^-$ (ring, H)		$^-O_3SCH_2CH_2N$ $NCH_2CH_2SO_3^-$ (ring)	

2.6   Choosing a Buffer

When you choose a buffer, you should keep in mind the following:

1. You need both HA and A$^-$ to have a buffer. As soon as you run out of one of them, you no longer have a buffer.

    You can think of a buffer system as a circle connecting the weak acid form and the weak base form (see Figure 2.1). If you add H$^+$ to the system, the system shifts to produce more of the weak acid form. If you add OH$^-$ to the system, the system shifts to produce more of the weak base form. Either way you eliminate the excess H$^+$ or OH$^-$.

    However, if you keep adding H$^+$, you eventually run out of one of them and no longer have a buffer. Therefore, the overall concentration of the buffer is important. If you have a buffer that is 0.1 M, you can add a lot more acid or base to it before it is exhausted than you can if you have a 0.001 M buffer.

2. The maximum buffering capacity is nearest the pKa of the buffer. Can you see why? At a pH equal to the pK_a, there are equal amounts of acid form and base form of the buffer present. This would be the best generic buffer, one that would be equally good at buffering against added acid or added base. Therefore, if you were trying to choose the best buffer, you would choose the one with the pK_a closest to the pH you wanted.

3. There is a usable range for a buffer. Given that you may not be able to get a buffer with a pK_a equal to the pH you need, you may have to settle for less than perfect. There are limits to how far you can stray from the pK_a however. How far away from the pK_a can

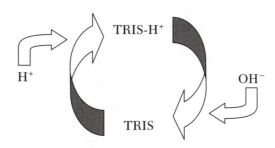

Figure 2.1 The circle of buffers

you go and still have a buffer? If your ratio of HA to A⁻ is 10, then effectively you have 91% of the molecule in the acid form and 9% in the basic form. This happens when your pH is one unit lower than the pK_a. At that point, there is not much of the base left, so you will not be able to buffer very well against added acid. Conversely, if your ratio of HA to A⁻ is 0.1, then 91% is in the basic form and only 9% is in the acid form. That means that there is not much acid left to buffer added base. This happens when the pH is one unit higher than the pKa. By convention, we say that a buffer is only effective at its $pK_a \pm 1$ pH unit, but the closer the pH is to the pK_a, the better.

Table 2.2 gives some of the common buffers you are likely to see along with their pK_as.

Table 2.2 Buffers and Their Properties

Compound	MW	pK_a at 20°C	$\Delta pK_a/°C$
ACES	182.2	6.90	
ADA, free acid	190.2	6.60	−0.011
ADA, sodium salt	212.2	6.60	−0.011
BES	213.2	7.17	−0.027
Bicine	163.2	8.35	−0.018
Boric acid	61.8	9.24	−0.018
CAPS	221.3	10.4	−0.009
CHES	207.3	9.5	−0.009
Citric acid	192.1	3.14	−0.009
	192.1	4.76	−0.009
	192.1	6.39	−0.009
Glycylglycine	132.1	8.4	−0.028
HEPES, free acid	238.3	7.55	−0.014
HEPES, sodium salt	260.3	7.55	−0.014
Imidazole	68.1	7.00	−0.014
MES, free acid	195.2	6.15	−0.011
MOPS, free acid	209.3	7.20	−0.006
PIPES, free acid	302.4	6.80	−0.009
Phosphoric acid (K_2HPO_4)	174.2	2.12	−0.009
	174.2	7.21	−0.009
	174.2	12.32	−0.009
TES, free acid	229.3	7.50	−0.020
Tricine	179.2	8.15	−0.021
Triethanolamine	185.7	7.66	−0.021
TRIS, (Trizma base)	121.1	8.30	−0.031
TRIS-HCl	157.6	8.30	−0.031

2.7   Effect of Concentration and Temperature

Effect of Temperature on Buffers

Buffers are at their best when the pH is adjusted correctly at the temperature at which they will be used. Most buffers are affected by temperature, some more than others. Table 2.2 shows the change in pK_a with temperature for some of the buffers. What this means is that if you calculate the change in pK_a with temperature, the pH would change the same amount, because changing the temperature will not change the ratio of base form to acid form. *Notice the sign on the number!* For example, if we raise the temperature of HEPES by one degree Celsius, we will lower the pH by 0.014.

Effect of Concentration on Buffers

Buffers are at their best when the pH is adjusted at the working concentration. However, most researchers make up concentrated solutions to save time and space. These solutions are then diluted to the appropriate concentrations before use. You will often see bottles labeled 20x TAE. This means that the solution is TAE buffer (TRIS, Acetate, EDTA) that must be diluted 20 to 1 before use. We have only talked about concentrations so far, and not about activities. pH is really a function of activity, a. The activity, a, is a function of the concentration, M, and the activity coefficient, γ. Activity coefficients change with concentration. The more charged the species is, the greater the change will be with changing concentration. The more exact equation for a weak acid would be

$$K_{HA} = \frac{\gamma_{H^+}\gamma_{A^-}[H^+][A^-]}{\gamma_{HA}[HA]}$$

Therefore, the K_a and pK_a will change with concentration even though the ratio appears to stay the same. Table 2.3 shows some typical changes.

What all this means to the pH of a buffer is rather complicated. Some buffers will show a pH increase with dilution; others will show a pH decrease. The zwitterionic buffers, such as HEPES, do not have as pronounced an effect compared to phosphoric acid or citric acid.

Table 2.3 Activity Coefficients at Different Concentrations			
Ion	**0.001 M**	**0.01 M**	**0.1 M**
H^+	0.98	0.93	0.86
OH^-	0.98	0.93	0.81
Acetate	0.98	0.93	0.82
$H_2PO_4^-$	0.98	0.93	0.74
HPO_4^{2-}	0.90	0.74	0.45
PO_4^{3-}	0.80	0.51	0.16

2.8 How We Make Buffers

We don't have to add both HA and A$^-$ to make a buffer. The easier way to do it is to titrate the HA with strong base or the A$^-$ with strong acid. If we start with a weak acid, such as acetic acid, and we add a strong base, such as NaOH, we will quantitatively force the production of A$^-$.

$$\textbf{HAc + NaOH} \longrightarrow \textbf{H}_2\textbf{O + Ac}^- + \textbf{Na}^+$$

This is not like a weak acid situation where we have to use the weak acid equation to figure out the extent of the reaction. Strong bases will win the tug of war against a weak acid any time. For every molecule of NaOH you put into the solution, one molecule of HAc will be converted to Ac$^-$.

Using the Henderson–Hasselbalch equation to figure out how much weak acid and salt to add to make a buffer is all well and good if we don't have a pH meter. Normally we *do* have one available, and there is a much easier way to make a buffer of known pH and concentration. Once you have the correct ratio of HA to A$^-$, it doesn't really matter how you arrived at that state.

If we want to make a phosphate buffer with a concentration of 0.1 M and a pH of 6, rather than calculating the ratios of $H_2PO_4^-$ and HPO_4^{2-} that we should add, we can just start out with an amount of $H_2PO_4^-$ and add sodium hydroxide until the pH is 6. Then by definition we have a buffer at the correct pH. The only trick is getting the concentration correct. In this example, if we wanted to end up with 100 mL of buffer, we would take 50 mL of 0.2 M NaH_2PO_4 and add NaOH until the pH was correct, then add water until the volume was 100 mL. We would then have the 0.1 M buffer that we need, and the pH would be 6.

2.9 The Big Summary

Summary of pH Equations

Definition	$pH = -\log[H^+]$
Water equation	$Kw = 10^{-14} = [H^+][OH^-]$
Henderson–Hasselbalch	$pH = pK_a + \log[A^-]/[HA]$
Weak acid only	$pH = \dfrac{pK_a - \log[HA]}{2}$
Weak base only	$pH = \dfrac{pKa + 14 + \log[base]}{2}$
Polyprotic intermediate	$pH = \dfrac{pK_a1 + pK_a2}{2}$

Note that in practice, diluting will affect the pH of a polyprotic acid solution, even though the concentration does not appear in our shortcut formula. For any solution, the pH should always be checked and readjusted after making the final dilution.

One Final Example Problem

◀▭▷ Practice Session 2.5

You take 2.38 g of HEPES in the acid form and bring the volume to 100 mL with water. We will call this solution A. You then add 3 mL of 1 M NaOH to make solution B. Finally, you add 7 more mL of 1 M NaOH to make solution C. What are the pH values for the three solutions?

The first thing to do is figure out how many moles you are dealing with. In this case, the MW of HEPES is 238.3 g/mol. Thus,

$$moles = 2.38/238.3 = 0.01 \text{ mol HEPES acid}$$

A. You dissolved that up to 100 mL, so you have a solution of 0.01 mol/0.1 L for a concentration of 0.1 M. This is a weak acid only situation, so you use the weak acid only equation:

$$pH = \frac{pK_a - \log[HA]}{2} = \frac{7.55 - \log(0.1)}{2} = \textbf{4.28}$$

B. Next, you add 3 mL of 1 M NaOH to the solution. Three mL of 1 M NaOH is 0.003 mol of OH^-, which will react with the weak acid to create weak base until you run out of one of them:

$$0.01 \text{ mol HA} + 0.003 \text{ mol OH}^- \longrightarrow 0.003 \text{ mol A}^- + 0.007 \text{ mol HA}$$

The 0.007 mol HA is what is left over after the base reacts. The 0.003 moles of A^- is what is formed when the hydroxide reacts.

You now have some HA and some A^-, so you have a buffer, and you use the Henderson–Hasselbalch equation:

$$pH = pK_a + \log(A^-/HA) = 7.55 + \log(0.003/0.007) = \textbf{7.18}$$

C. Now you add 7 mL more of NaOH. That would be another 0.007 mol of OH^-. How handy! That happens to be the exact amount of weak acid you had left over from the last part. Thus, you will now use up all of the weak acid and have only 0.01 mol of the base form of HEPES left. This is a weak base only situation.

$$pH = \frac{pK_a + 14 + \log[A^-]}{2} = \frac{7.55 + 14 + \log(0.01/0.110)}{2} = \textbf{10.25}$$

For this equation, the only trick is to remember that it calls for the concentration of the weak base. Note that the volume has changed during the experiment. That is why we divided the 0.01 mol by 0.11 L, because the new volume is the original volume of 100 mL (0.1 L) *plus* the total volume of base added (10 mL or 0.01 L).

This problem can appear in slightly different forms where we start with the weak base and go toward the acid or where we start with a buffer and work back. The principles are the same.

2.10   Why Is This Important?

If you work in the sciences, you will use buffers every day. Many people use them blindly and make mistakes with them. This is even true of graduate students and postdocs. If you understand this lab, you will be able to choose the correct buffer for the job and understand how it works. You will know whether it will work well at different temperatures and concentrations. You will know the fastest way to make your buffer and be able to choose the best starting compounds. Making buffers is far from glamorous, but you will do it so many times that understanding it will save you time and money.

Tip 2.3

> If you pull a bottle of Trizma base off the shelf and read the label, you will see that it says the pK_a is 8.3. If you then take a certain amount of the compound and add it to water and do nothing else, the pH will most certainly *not* be 8.3. It will be very much higher than 8.3. You only get the pH to equal the pK_a if you titrate the Trizma base with strong acid until you have added half of an equimolar amount of acid.

2.11   Expanding the Topic

Converting Ratios to Percentages

Many students find it difficult to calculate the percentage of a buffer that is in the weak acid or weak base form. This is usually because they have difficulty converting the ratios to percentages. For example, if you use the Henderson–Hasselbalch (H–H) equation and calculate that the ratio of A^-/HA is 1, then you would have no difficulty saying that the buffer is 50% A^- and 50% HA. If there were a total of 0.02 mol of buffer, there would be 0.01 mol in the A^- form and 0.01 mol in the HA form.

What if the ratio is not so easy as that? Let's say you use the H–H equation and find that the ratio is 2. That means that for every HA, there are 2 A^-. That would give you percentages of 66% for A^- and 33% for HA. Whenever you have a ratio, such as 2/1, there are a total of 3 parts in the system. The numerator is 2/3, and the denominator is 1/3. To get the percentage, you divide the numerator or denominator by the total.

If you did the H–H equation and the ratio was 4, then that means A^-/HA is 4 to 1. There are a total of 5 parts in the system, so the A^- is 4/5, which is 80%.

Even if the ratio comes out to be really weird, like 4051 to 1, you would still calculate the percentages the same way. If the ratio of A^-/HA is 4051 to 1, then there are a total of 4052 parts. The percentage of the buffer in the A^- form would be 4051/4052 or 99.975%.

History of the Shortcut Formulas

To determine the exact pH of a solution of weak acid, it is necessary to do some algebra that includes a quadratic equation. If we have a weak acid, HA, it will dissociate in water to give H^+ and A^- in equal amounts:

$$HA \rightleftharpoons H^+ + A^-$$

If the initial concentration of HA is 0.1 M, then after it dissociates, the concentration will be reduced by the amount that has dissociated. If we say that an unknown amount of H^+ will dissociate and we call that amount x, then the new concentration of HA will be $0.1\ M - x$. Remember that the acid dissociation expression is as follows:

$$K_a = \frac{[H^+][A^-]}{[HA]}$$

If we designate the amount of H^+ to be x, then the amount of A^- must also be x, because they are formed in equal amounts from the dissociation. The equation would then be as follows:

$$K_a = \frac{x^2}{0.1 - x}$$

This is a quadratic equation. To determine x, and therefore the $[H^+]$, we would need to solve for x in the equation. Most calculators can do this easily nowadays. However, if we make one simplifying assumption, this equation becomes much easier. Let's assume that the amount, x, is small compared to the initial concentration of HA. When would this happen? When the acid is weak, which is precisely when we are using this equation in the first place.

If we assume that the x is negligible when compared to the [HA], then the equation simplifies to the following:

$$K_a = \frac{[H^+][A^-]}{[HA]} = \frac{x^2}{[HA]} = \frac{[H^+]^2}{[HA]}$$

Solving for $[H^+]$ would give us the following equation:

$$[H^+] = \sqrt{K_a[HA]}$$

To obtain an expression involving pH, we put the expression into logarithmic form:

$$\text{Log}[H^+] = \tfrac{1}{2}\log K_a[HA] = \tfrac{1}{2}(\log K_a + \log[HA])$$

$$-\log[H^+] = \frac{-\log K_a - \log[HA]}{2} = \frac{pK_a - \log[HA]}{2} = pH$$

Similar derivations can be done for the weak base solution and the Henderson–Hasselbalch equation (see *Biochemical Calculations*, by I. Segel). Because this equation is based on an assumption, anytime the assumption is wrong, so is the calculation. The stronger the acid, the more the real pH will deviate from that calculated by this formula.

Experiment 2

Preparation of Buffers

In this experiment, you will learn how to choose and prepare a buffer. You will then see how the pH of the buffer responds to dilution and compare how buffered and unbuffered systems respond to addition of acids and bases.

Prelab Questions

1. Calculate the weight of the buffers you would use to make the buffers for part A for all of the buffer possibilities listed under procedures part A, step 1. In other words, how many grams do you need to make 100 mL of a 0.1 M buffer?

2. If we give you HEPES in the basic form and ask you to make a buffer of pH 8.0, will you have to add HCl or NaOH? Why? (With commercial buffers, there is always an acid form and a basic form that one can buy. It is not obvious from the name of the compound, so one has to look to see if it is acid or basic. If HEPES is bought in the acid form, then one could write the equation:

$$HEPES \rightleftharpoons HEPES^- + H^+$$

Objectives

Upon successful completion of this laboratory, the student will be able to

1. Calculate the pH of solutions of strong acids or bases, weak acids or bases, buffers, and/or combinations of these.

2. Explain how a buffer resists change in pH.

3. Prepare an appropriate buffer for a given pH.

4. Calculate the theoretical pH of the buffer after adding a known quantity of acid or base.

5. Predict the effect of changing temperature and concentration on buffer pH.

6. Standardize and operate a pH meter.

Experimental Procedures

Materials

Solid buffers

Standard buffer, pH 4

Standard buffer, pH 7

NaOH, 1 M

HCl, 1 M

pH meters

Volumetric flasks

Methods

Part A: Preparation of Buffers

For this first part, you will make two buffers starting with the solid material. This is the most common way to make buffers. You will be given a desired pH, and your task is to prepare 100 mL of two appropriate buffers at a concentration of 0.10 M. One of the buffers will be a phosphate or citrate buffer, and the other will be one of the others.

1. Using the following table, choose the most appropriate buffer compounds for your pHs. Proceed with steps 2–7 for both buffers.

Buffer	pK_a1	pK_a2	pK_a3	Formula Weight (g/mol)
Acetate	4.76			136.1
CAPS	10.4			221.3
Citrate	3.06	4.74	5.40	294.1
HEPES	7.55			238.3
Phosphate	2.12	7.21	12.32	142.0
Tricine	8.15			179.2
TRIS	8.3			121.1
_____	_____			_____
_____	_____			_____

2. Calculate the weight of the buffer you would need to make 100 mL of a 0.100 M solution. Weigh out the correct amount and dissolve in 50 mL water.

3. Standardize the pH meter at pH 4,7, and 10. Set up the beaker with your buffer solution on a stirplate such that you can stir the solution and read the pH continuously. If you have no stirplate, just swirl the beaker often while you add acid or base.

4. Use 1 M NaOH or 1 M HCl to titrate to the desired pH. Add the acid or base a drop at a time. By doing this, you effectively change some of the acid form to the base form or vice versa until the ratio is the correct one to give you the pH you want.

5. Add water until the volume is about 99 mL.

6. Recheck the pH to make sure it has not changed. If it has, correct it with the NaOH or HCl. **Warning!** *You might want to use a lower concentration of NaOH or HCl.*

7. Bring the volume to 100 mL and save this solution for later.

Part B: Effect of Concentration on pH

For this part, you must have a digital pH meter.

1. Take 10 mL of each of your two buffers and dilute with deionized water to give a final concentration of 0.01 M. Save these solutions.

2. Take 10 mL of your diluted buffers from step 1 and dilute to a concentration of 0.001 M. Save these solutions.

3. Measure the pH of the undiluted and diluted solutions.

Part C: pH Measurement of Other Solutions

Measure the pH of the following solutions.

Distilled water

Unknown #_____

Unknown #_____

Part D: Why a Buffer Is a Buffer

1. Put 50 mL of one of your original 0.1 M buffers in a beaker. If your buffer has a pH higher than its pK_a, add 0.5 mL of 1 M HCl to it. Record the new pH. If your buffer has a pH lower than its pK_a, add 0.5 mL of 1 M NaOH to it. Record the new pH.

2. Repeat step 1 but use 50 mL of water instead of your buffer. Add either acid or base depending on what you did in step 1.

Analysis of Results
Experiment 2: Buffers

Data

Part A

Buffer 1: _____ weight (g): _____ Original pH: _____

Buffer 2: _____ weight (g): _____ Original pH: _____

Part B

Buffer 1   pH of 0.1 M _____   pH of 0.01 M _____   pH of 0.001 M _____

Buffer 2   pH of 0.1 M _____   pH of 0.01 M _____   pH of 0.001 M _____

Part C

Distilled water pH: _____

Unknown _____   pH: _____

Unknown _____   pH: _____

Part D

Buffer chosen: _____   pH: _____   pK_a: _____

Acid or base added: _____

pH after adding acid or base: _____

pH of 50 mL of water: _____

pH after adding acid or base: _____

Calculations

1. If you added 3 mL of 1 M NaOH to your 100 mL of buffer, would it still be a usable buffer according to our conventions? Explain why or why not (math would be good here.) You may do this problem on one of your buffers. Whichever one you choose, your lab partner, if any, must choose the other.

2. Calculate the theoretical pH of one of your buffers at 0° C. Assume that room temperature is 22°C. If none of your buffers was listed on the table of changing pK_a with temperature, do this problem for TRIS at pH 8.0.

3. What would be the most efficient way to make up a HEPES buffer at pH 8.5? What starting compounds and reagents would you choose to use?

4. When Dr. Farrell was a graduate student, he once made up a pH 8.0 sodium acetate buffer. Why would the casual observer of this buffering faux pas come to the conclusion that he had the intellectual agility of a small soap dish?

5. If you made up a solution of 100 mL 0.1 M TRIS in the acid form, what would be the pH?

6. If you added 3 mL of 1 M NaOH to the solution in 5, what would be the pH?

Additional Problem Set

1. Calculate the pH of a 0.1 M HCl solution.

2. Calculate the pH of a 0.1 M NaOH solution.

3. What is the concentration of $[H^+]$ in M, mM, and μM for a solution of pH 5?

4. If you mix 10 mL of a 0.1 M HCl solution with 8 mL of a 0.2 M NaOH solution, what will be the resulting pH?

5. If a weak acid, HA, is 3% dissociated in a 0.25 M solution, calculate the K_a and the pH of the solution.

6. What is the pH of a 0.05 M solution of TRIS acid ($pK_a = 8.3$)?

7. What is the pH of a 0.045 M solution of TRIS base?

8. If you mixed 50 mL of 0.1 M TRIS acid with 60 mL of 0.2 M TRIS base, what would be the resulting pH?

9. If you added 1 mL of 1 M NaOH to the solution in problem 8, what would be the pH?

10. How many total mL of 1 M NaOH could you add to the solution in problem 8 and still have a good buffer (i.e., within one pH unit of the pK_a)?

11. If you were making 100 mL of a 0.1 M HEPES buffer starting from HEPES in the basic form, would it be prudent to go get 50 mL of 1 M HCl from the community reagent bottle to use for your titration?

12. An enzyme-catalyzed reaction is carried out in 100 mL of a solution containing 0.1 M TRIS buffer. The pH of the reaction mixture at the start was 8.0. As a result of the reaction, 0.002 mol of H^+ were produced. What was the ratio of TRIS base to TRIS acid at the start of the experiment? What was the ratio at the end of the experiment? What was the final pH?

13. The pK_a of HEPES is 7.55 at 20°C and its MW is 238.31. Calculate the amounts of HEPES in grams and of 1.0 M NaOH in milliliters that would be needed to make 300 mL of 0.2 M HEPES buffer at pH 7.2.

Webconnections

For a list of Web sites related to the material covered in this chapter, go to **Webconnections** at the *Experiments in Biochemistry* site on the Saunders College Publishing Web page. You can access this page at http://www.saunderscollege.com. **Webconnections** are under *Experiments in Biochemistry* in the Biochemistry portion of the Chemistry page.

References and Further Reading

R. F. Boyer, *Modern Experimental Biochemistry*, Addison-Wesley, 1993.
M. Campbell, *Biochemistry*, Saunders College Publishing, 1998.
P. A. Ciullo, *Bicarb: Buffers, Bubbles, Biscuits, and Earthburps*, Maradia Press, 1997.
R. Cecil Jack, *Basic Biochemical Laboratory Procedures and Computing*, Oxford University Press, 1995.
C. Mohan, *Buffers: A Guide for the Preparation and Use of Buffers in Biological Systems*, Calbiochem Co., 1995.
J. F. Robyt and B. J. White, *Biochemical Techniques*, Waveland Press, 1990.
I. H. Segel, *Biochemical Calculations*, 2nd ed., Wiley Interscience, 1976.
J. Stenesh, *Experimental Biochemistry*, Allyn and Bacon, 1984.

Chapter 3

Spectrophotometry

Topics

Introduction

This chapter deals with the most often used theories and techniques found in a biochemistry lab, those of spectrophotometry. Almost every experiment we do in lab involves the use of a spectrophotometer in some way. Along with using a Pipetman, how you use a spectrophotometer will affect the results you get with your experiments. If you learn to use them well, your labs will run smoothly. As with any piece of equipment, the old adage "garbage in, garbage out" is very true with even the simplest spectrophotometer.

3.1 Absorption of Light

Almost all biochemical experiments eventually use spectrophotometry to measure the amount of a substance in solution. Spectrophotometry is the study of the interaction of electromagnetic radiation with molecules, atoms, or ions.

Light or electromagnetic radiation has a wave and particle nature. The **wavelength, λ,** of light is the distance between adjacent peaks in the wave. The **frequency, ν,** is the number of waves passing a fixed point per unit of time (see Figure 3.1.)

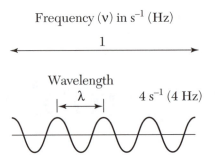

Frequency (ν) in s⁻¹ (Hz)

Figure 3.1 Wave nature of light

These parameters can be further defined by the equation

$$\lambda = c/\nu$$

where c is the speed of light.

Photons of different wavelength have different energies. These energies can be calculated by the equation

$$E = hc/\lambda = h\nu$$

where h is Planck's constant. Therefore, the longer the wavelength, the less energy the light has and vice versa.

Figure 3.2 shows the relationship between the wavelength of light and the common types of electromagnetic radiation. As you can see, those regions where the wavelength is very short correspond to the types of radiation that you know are powerful and often harmful, such as X rays, γ-rays, and ultraviolet radiation.

Most compounds have a certain characteristic wavelength or wavelengths of light that they absorb. This process can be diagrammed as in Figure 3.3. Thus the solution looks green to us because green light (blue and yellow) is **transmitted** while the red light is **absorbed.**

A solution may contain many compounds that absorb at many different wavelengths, but

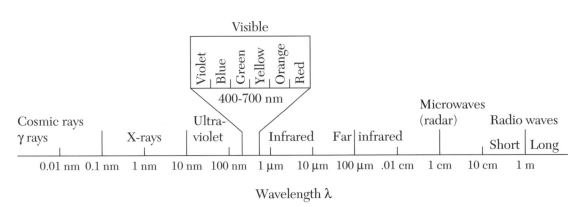

Figure 3.2 Wavelength regions of light

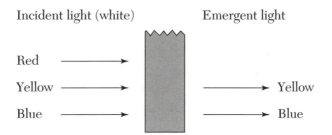

Figure 3.3 Absorption of light by a solution

if a compound we are interested in absorbs at a unique wavelength, we can determine its concentration even in a solution of other compounds.

3.2 Beer–Lambert Law

If a ray of monochromatic light (1 wavelength) of initial intensity I_0 passes through a solution, some of the light may be absorbed, so that the transmitted light I is less than I_0. (See Figure 3.4.) The ratio of intensities I/I_0 is called the **transmittance** and is dependent on several factors.

1. If the concentration, c, of the absorbing solution increases, then the transmittance will decrease.

2. If the pathlength, l, that the light must travel through increases, then the transmittance will decrease.

3. If the nature of the substance changes or another substance that absorbs more strongly is used, then the transmittance will change. The nature of the substance is reflected in ϵ, the extinction coefficient, also called the absorptivity constant.

   An equation can be written that incorporates these ideas:

$$\log I_0/I = \epsilon lc$$

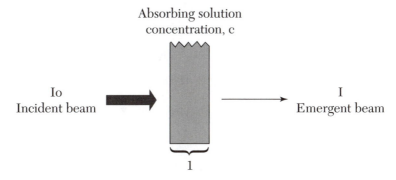

Figure 3.4 Relationship between I, I_0, l, and c for a solution absorbing monochromatic light

where I_0 = intensity of incident light
   I = intensity of transmitted light
   ϵ = extinction coefficient
   l = path length through solution
   c = concentration of absorbing solution

log I_0/I is usually called the **absorbance** and is abbreviated A.
   This law:

$$A = \epsilon lc$$

is called the Beer–Lambert law (Beer's law).

Some Points to Consider

1. Absorbance, A, has no units. It is just a number that you can read off of the spectrophotometer. The wavelength is often specified along with the absorbance, such as $A_{540} = 0.3$.

2. The extinction coefficient, ϵ, is the absorbance of a unit solution and has units of reciprocal concentration and path length. The most common ϵ values recorded are for a path length of 1 cm and a 1 M solution. Therefore the expression $\epsilon 600 = 4000 \text{ M}^{-1}\text{cm}^{-1}$ means that a 1 M solution would have an absorbance at 600 nm of 4000 if you use a cuvette with a diameter of 1 cm.

3. Remember that the pathlength, l, is usually in cm and if not specified is assumed to be 1 cm.

4. The concentration, c, will have units that are the reciprocal of the units for ϵ.

5. You need at least 2 mL in a cuvette in order to read it with the standard Milton Roy Spectronic® 20 spectrophotometer.

6. Lots of things can interfere with your use of a spectrophotometer. If the cuvette is smudged or scratched, light will be scattered by the tube rather than absorbed by the solution. If you do not have sufficient volume (see Point 5) the light may pass over the solution instead of going through it. The spectrophotometer must be well calibrated before use.

If a substance obeys the Beer–Lambert law, then a plot of A vs. c is straight, as in the ideal line in Figure 3.5. More often, however, the line is curved, shown as reality in Figure 3.5.

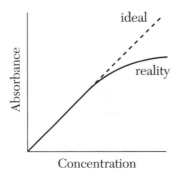

Figure 3.5   Absorbance vs. concentration for substances obeying Beer's law

➡ Practice Session 3.1

We measure the absorbance of a solution of compound X, which absorbs at 540 nm. The cuvette has a width of 1 cm, the extinction coefficient at 540 nm is 10,000 M^{-1} cm^{-1}, and the absorbance is 0.4. What is the concentration of compound X?

From Beer's law:

$$A = \epsilon l c$$

$$c = A/\epsilon l = \frac{0.4}{}$$

$$= 4 \times 10^{-5} M$$

Reagent Blanks

A reagent blank is a control in which everything is included except the substance for which we are testing. One problem often encountered in spectrophotometry is that there is an absorbance at a given wavelength not due to the substance of interest. We handle that by mixing up all of the solutions in a tube except that substance and then reading the absorbance. The absorbance of the reagent blank is then subtracted from the other readings.

➡ Practice Session 3.2

If we want to read the absorbance at 595 nm of a protein solution mixed with the colorizing reagent called Bradford reagent, what would be a suitable reagent blank?

We want everything that might contribute to an absorbance at 595 nm except the protein we are trying to measure. Therefore the best reagent blank would be a tube of Bradford reagent without any added protein. By zeroing the machine on this tube, the absorbance due to the Bradford reagent is subtracted out automatically.

ESSENTIAL INFORMATION

Beer's law allows you to calculate the concentration of a substance in solution after measuring the absorbance with a spectrophotometer. Before you use a spectrophotometer, it must be properly calibrated or zeroed. If it is not, the numbers you generate will be meaningless.

As we shall see in later chapters, you can also measure a changing concentration, such as the product of an enzymatic reaction, by measuring the changing absorbance. Beer's law only works if you know that the relationship between absorbance and concentration is linear. This is not always the case. If you cannot use Beer's law, you proceed to the next technique, which is making a standard curve.

3.3 Standard Curves

Figuring out the concentration of a substance as in Practice Session 3.1 is all well and good *if* you know the extinction coefficient and *if* you know that the system obeys Beer's law at that concentration. When these things are not known, which is most of the time, a standard curve is prepared. A standard curve is a plot of A vs. a varying amount of a substance. Then, an unknown concentration can be determined from the graph.

☞ Practice Session 3.3

If we have a compound X of varying concentrations in a phosphate buffer, pH 7.0, and the absorbencies are as follows,

Tube #	Concentration (mM)	Absorbance	Corrected Absorbance
1	0	0.05	0.00
2	1	0.15	0.10
3	2	0.25	0.20
4	3	0.35	0.30
5	4	0.45	0.40
6	5	0.55	0.50

what is the concentration of a sample of X if the absorbance equals 0.30?

First, you must understand **corrected absorbencies.** Tube 1 has an absorbance of 0.05, but it does not contain any of the compound we are measuring. This tube is our reagent blank, and it has an absorbance. We want our graph to have a curve that goes through zero. There are two ways of dealing with this. The first is to subtract the absorbance (0.05) from all of the absorbencies. This will give the data shown in the last column. The second is to zero the spectrophotometer with tube 1. That way, the subtraction is done automatically. We will always use corrected curves, even though the difference between a corrected curve and an uncorrected one is largely cosmetic.

Continuing with the example, first we subtract the reagent blank (tube 1) from the rest to give the corrected absorbance; then we plot A vs. c (corrected) (see Figure 3.6). On the graph, we look for the A that corresponds to the unknown. $0.30 - 0.05$ (blank) = 0.25. From the graph, $A = 0.25$ corresponds to the concentration of 2.5 mM, which is the answer.

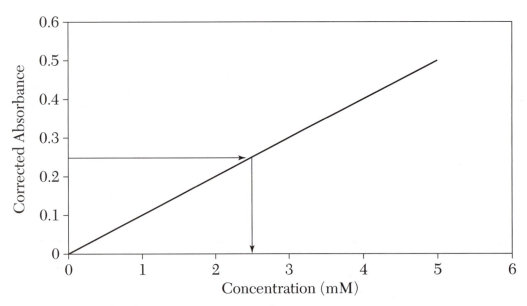

Figure 3.6   Corrected absorbance vs. concentration for Practice Session 3.3

3.4 Protein Assays

One of the most common uses for spectrophotometry, which also happens to use standard curves, is the protein assay. Many biochemical studies at some point require the knowledge of how much protein you have in a sample.

Ultraviolet Absorption

Proteins can be assayed easily if you have a spectrophotometer that can measure light in the ultraviolet region. The amino acids tryptophan and tyrosine absorb strongly at 280 nm. This fact allows the scientist to scan for proteins at this wavelength. This is often done as you purify a protein with some chromatographic technique. However, if you want to get quantitative you would have to know the exact ϵ_{280} for the protein. If the protein contains few aromatic residues or the extinction coefficient is low, the ultraviolet method would not be suitable. Also, many spectrophotometers found in teaching labs do not have UV capability.

Colorizing Reagents

Many assays have been developed over the years that can compensate for the inability to use ultraviolet absorption. Most of them depend upon certain dye molecules that react with parts of the protein to give a colored complex that can then be measured. Once you have a colored complex, you can use visible light spectrophotometry.

One of the most common and easiest to use is the Bradford method. This method uses a dye called Coomassie Brilliant Blue G-250, which has a negative charge on it. The dye normally exists in a red form that absorbs light maximally at 465 nm. When the dye binds to the positive charges on a protein, it shifts to the blue form, which absorbs maximally at 595 nm. Many proteins have the same response curve to this dye, making the Bradford method reproducible between many experimenters. It is also very rapid. The reaction occurs in a couple of minutes, and the colored product is stable for over an hour. In addition, the protocol calls for a protein sample of up to 100 μL to be added to 3–5 mL of Bradford reagent. With such a large difference in volumes between the sample and the protein reagent, it is not necessary to bring all samples up to the same 100 μL volume before reagent addition. This saves time in setting up the assays.

ESSENTIAL INFORMATION

To make and use a standard curve, you set up a series of tubes with varying amounts of substance in them. After measuring the absorbance, you can plot the corrected absorbance vs. the amount you put in. Sometimes you will plot absorbance vs. concentration, but the most useful value to plot is either a weight (such as micrograms or milligrams) or an amount in a molar-based unit, such as micromoles, millimoles, and so on. Once you have your standard curve, you can use it to determine the concentration of an unknown. The absorbance value of the unknown tube must fall within the line of your standard curve, preferably within the linear region. You may not extrapolate your line beyond the highest concentration standard you have.

3.5 Why Is This Important?

No single piece of equipment is used more in biochemistry or any life science than the spectrophotometer. We measure almost everything using one. Granted, most research labs have fancier ones that do more of the work for you, but at the heart of it is a simple light source, a prism or grating for controlling the wavelength, and a sample holder. It is far too easy to get caught up in the fun of pushing buttons without understanding how the machine works. With any piece of equipment, the numbers generated only have meaning if the machine was calibrated and used properly. In most undergraduate labs, you will use the Spectronic 20, Spectronic 20 D, Spectronic 20 D+, or Spectronic 21 constantly. Take the time to really learn how to use them now. The payoff will be great.

> **Tip 3.1**
>
> There are a few tricks to getting good results with a Spectronic 20.
>
> First, zero the machine often and close to when you are taking your measurements.
>
> Second, make sure you have over 2 mL in the cuvette.
>
> Third, be careful with the number of cuvettes you use. The best data come from the fewest number of cuvettes.

3.6 Expanding the Topic

Calculating Concentrations from Graphs

A major source of frustration to many students is figuring out just what to plot against absorbance for these types of experiments. There are two quantities that could be plotted. The first is the final concentration of the product in the reaction vessel in moles per liter, millimoles per liter, milligrams per milliliter, percent, and so on. The second is the amount of product put into the reaction vessel in milligrams, grams, millimoles, moles, and so on. The latter is often more useful because what we want to determine is the concentration of the unknown solution *before* it was put into the reaction vessel. As an example, consider the determination of creatinine. The objective is to determine the [creatinine] in the unknown serum. A hypothetical table might look like this:

Reagent/Tube	1	2	3	4	5	6
1 mg/mL creatinine	0	0.5	1.0	2.0	3.0	—
Water	3	2.5	2.0	1.0	0	2.0
Serum	—	—	—	—	—	1.0
Picrate	3	3	3	3	3	3
A_{500}	0	0.1	0.2	0.4	0.6	0.5

For simplicity let's give the blank an A of zero (we should be so lucky). Now we will calculate the concentration of the unknown by the two methods. Keep track of the steps involved.

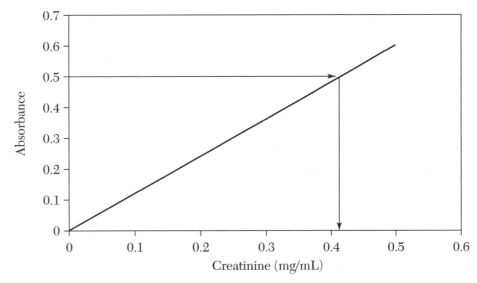

Figure 3.7 Absorbance vs. [creatinine]

Method 1 (not recommended)

We will plot A vs. final [creatinine], so we need to calculate the concentrations. For each tube, 1 mg/mL creatinine × mL added = mg creatinine added. Added mg creatinine/6 mL = final concentration.

Tube	1	2	3	4	5	6
mg creatinine	0	0.5	1	2	3	—
mg/mL final	0	0.083	0.167	0.333	0.5	—

Then we plot A vs. [creatinine] as in Figure 3.7.

From the graph we find that $A_{500} = 0.5$ gives us a concentration of 0.417 mg/mL. Each tube has 6 mL, so the result is

$$0.417 \text{ mg/mL} \times 6 \text{ mL} = 2.5 \text{ mg creatinine}$$

The 2.5 mg creatinine came from 1 mL, so the concentration is

$$2.5 \text{ mg/1 mL} = 2.5 \text{ mg/mL}$$

With this method, at one point you divided by 6 mL and then turned right around and multiplied by 6 mL a few steps later.

Method 2 (recommended)

This time we will plot A_{500} vs. mg creatinine added, as in Figure 3.8:

Tube	1	2	3	4	5	6
mg Creatinine	0	0.5	1	2	3	—

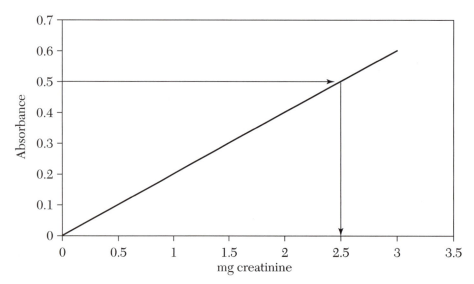

Figure 3.8 Absorbance vs. mg creatinine

From the graph we find that $A_{500} = 0.5$ gives us 2.5 mg creatinine added. Because it came from 1 mL, we again calculate that we have 2.5 mg/mL creatinine in the serum.

I hope you can see that this method eliminated the needless conversion to concentration within the reaction vessel and then back again. To help you remember things like this, ask yourself, "What am I trying to figure out?" With assays designed to determine the concentration of something, the concentration in the assay tube is usually irrelevant. It is the concentration in the original sample you want to know.

3.7   Tricks of the Trade

Choosing Test Tubes

How can you tell what test tubes to use? Any time you are going to be vortexing solutions in tubes, you want a tube that will be half full or less. Otherwise, your vortexing will be ineffective at best and messy at worst.

When should you use just one cuvette? In a standard curve for a protein assay and other similar assays, it is best to set up the reagents in large tubes and then pipet or pour some into one cuvette. This will accomplish two things. First, it keeps you from trying to mix up the reagents in tubes that are too small and too expensive to use for that purpose. Second, it is more accurate to read the same tube over and over, so that differences in the tubes do not confuse differences in the solutions. In other words, *you want as few variables as possible!*

Choosing Pipets

Pipets are at their best when used to capacity. You would not want to pipet 1 mL with a 10-mL pipet. You would have a huge error. There is error any time you pipet. The error grows as you use less and less of the capacity of the pipet. It also grows as you use a pipet again and again. For instance, you would not want to pipet 5 mL by using a 0.2-mL pipet 25 times.

For the experiment you will do in this chapter, you have a choice of weapons for pipetting the BSA into the tubes. The preferred ones are the Hamilton syringe and the disposable (dispo) micropipet. The Hamilton is the most accurate pipet available. There are also micropipets of 5, 10, 20, 50, and 100 μL. The advantage of the Hamilton syringe is its accuracy, but the disadvantage is that you have to clean it. Also, if you bend the needle or the plunger, it is broken for good. A Pipetman can also be used. The advantages are the replaceable tips. The disadvantage is that they need to be calibrated often and are not very accurate in untrained hands.

Experiment 3

Beer's Law and Standard Curves

In this experiment, you will learn how to use a spectrophotometer to calculate concentrations using Beer's law. You will also do a simple protein assay and learn to create standard curves for the determination of the protein concentration in an unknown.

Prelab Questions

1. If we give you 1 mL of NADH solution for part A, what is the minimum dilution you must make to do the experiment? (*Hint:* this refers to the amount of solution you need to have in a cuvette in order to use it in a spectrophotometer).

2. If you add 1 mL of water to 2 mL of NADH, mix, and get an absorbance of 0.2, what was the concentration of the original NADH solution?

3. What size test tubes will you use to mix the reagents for the Bradford protein assay?

Objectives

Upon successful completion of this experiment, the student will be able to

1. Zero the spectrophotometer at a variety of wavelengths and measure the absorbencies of solutions.

2. Decide when dilutions must be made and make the appropriate dilutions.

3. Calculate absorbencies, concentrations, and extinction coefficients using Beer's law.

4. Make standard curves and determine concentrations from them.

5. Define a reagent blank and decide what reagents must be present in one.

6. Design protocols for the creation of a standard curve.

Experimental Procedures

Materials

Spectrophotometers and cuvettes

NADH unknowns

Bradford protein reagent

BSA, 1 mg/mL

BSA, unknown

Methods

Part A: Using Beer's Law to Determine Concentration

In this part of the experiment, you will use Beer's law to determine the concentration of a solution of NADH, a reagent that we will use later in the course when we do enzyme purification. NADH has an extinction coefficient of 6220 M^{-1} cm^{-1} at 340 nm. The path length for the cuvette is 1 cm.

1. Obtain a solution of NADH of unknown concentration.

2. Warm up and zero the spectrophotometer at 340 nm or at the wavelength as close to 340 nm that you can. Older machines cannot be zeroed at 340 nm, but can be zeroed somewhere between 340 and 360.

3. Make a minimal dilution of the NADH to give you enough solution to measure in the cuvette.

4. Measure the absorbance of the NADH. If the absorbance is greater than 0.8, you will have to dilute the solution with water and remeasure. Make sure you record these dilutions. You need to know how much NADH you added to how much water.

5. Use the absorbance and any dilutions you made to determine the millimolar concentration of the NADH.

Part B: Setting Up a Standard Curve

In this part you will set up a standard curve for a protein determination, called the Bradford method. The protein standard will be bovine serum albumin, BSA, which is a generic protein that most people use for their protein assays. We usually set up a series of tubes with varying amounts of BSA and a constant amount of Bradford reagent. By plotting the number of milligrams or micrograms of BSA on the x-axis and the corrected absorbance on the y-axis, we can determine the concentration of unknown proteins from the graph.

1. Warm up the spectrophotometer at 595 nm.

2. Set up ten large, clean test tubes to use for the assays. As a general rule, it is better to use the big tubes for the reactions, then pour a couple of milliliters of the solution into a cuvette-sized tube to read it, rather than setting up the reaction in the cuvettes. The cuvettes are too small to mix most reactions in, and then you risk permanently discoloring the cuvettes.

3. Set up a protocol as in Table 3.1. Using a dispo micropipet or a Hamilton syringe, pipet the BSA standard into the tubes. This would be a particularly bad time to use a Pipetman. The protein concentration is very high and the volume is low, so any pipetting error will lead to poor standard curves. In most people's hands, a Pipetman is not accurate enough.

4. Obtain an unknown BSA solution. Choose a volume of the unknown to assay and pipet into tube 9. This volume must be 100 μL or less.

Table 3.1 Protocol for Protein Determination

Reagent/Tube (mL)	1	2	3	4	5	6	7	8	9
BSA standard 1 mg/mL (μL)	0	10	20	30	40	50	75	100	0
Unknown protein	0	0	0	0	0	0	0	0	?
Bradford reagent	3.0 mL								

5. Add 3 mL of Bradford reagent to each tube. Vortex immediately after adding the reagent to each tube. Do not wait until you have added it to all tubes.

6. Let the tubes sit about 10 min before reading the absorbencies. Once the color develops, it is stable for over an hour.

7. Make a plot to determine how much BSA you can add without the curve straying from linear. You may find that it was linear through your highest standard (100 μL), but it may also veer off at a lower volume of standard. If the absorbance of the unknown BSA is higher than your highest standard point that is on the straight part of the curve, you will need to make up another tube using less of the unknown BSA. You also want the corrected absorbance of your unknown to be 0.8 or less.

Analysis of Results

Experiment 3: Beer's Law and Standard Curves

Data

Part A: Beer's Law

1. Unknown # _____

2. Dilution of NADH _____

3. Describe how you arrived at this dilution:

4. Absorbance of NADH _____

Part B: Bradford Protein Assay

1. Unknown # _____

2. Fill in the following table for your raw data:

Reagent/Tube	1	2	3	4	5	6	7	8	9	10	11
BSA Volume (μL)	0	10	20	30	40	50	75	100			
Protein (μg)											
Absorbance (corr)											

Analysis of Results

Part A: Beer's Law

1. Concentration of NADH _____

2. Describe how you arrived at this number.

Part B: Bradford Protein Assay

1. Make a proper graph of corrected absorbance vs. μg protein and attach it to this report.

2. Calculate the number of micrograms of protein in your unknown assays.

3. Calculate the concentration of the protein unknown. This is done by dividing the weight of protein you determined in step 2 by the volume of sample you put into the Bradford reagent.

    Unknown concentration = _____ μg ÷ _____ μL = _____ mg/mL

Questions

1. What is the theoretical absorbance at 340 nm of a 0.1 M solution of NADH, assuming a pathlength of 1 cm?

2. What dilution would be necessary to get the absorbance from #1 down to 3.1?

    (_____ mL 0.1 M NADH to _____ mL H_2O)

3. Absorbance at 340 nm of a 0.02 mM NADH solution is 0.124 with a 1-cm pathlength. What would be the absorbance with a 1.2-cm pathlength?

4. Five μL of a 10/1 dilution of a sample were added to 3 mL of Bradford reagent. The absorbance at 595 was 0.78, and according to a standard curve corresponds to 0.015 mg protein on the x-axis. What is the protein concentration of the original solution?

Experiment 3a:

Protein Concentration of LDH Fractions

In this experiment, you will learn to do the Bradford protein assay, a rapid and simple assay for protein concentration. This will allow you to calculate the protein concentration of the LDH fractions from your purification experiments, which will, in turn, allow you to calculate the specific activity and the fold purification at each step.

Prelab Questions

1. Why do you want to use large test tubes for the Bradford assay?

2. With the standard Bradford assay, why is it not necessary to equalize all the sample volumes before adding the Bradford reagent?

3. Why is it not necessary to dilute your protein samples with buffer?

Objectives

Upon successful completion of this experiment, the student will be able to

1. Make standard curves and determine protein concentrations from them.

2. Design protocols for the creation of a standard curve.

3. Complete the LDH purification table for Chapter 7.

Experimental Procedures

Materials

Spectrophotometers and cuvettes

Bradford protein reagent

BSA, 1 mg/mL

LDH fractions from purification

Methods

1. Turn on the spectrophotometer and set the wavelength to 595 nm.

2. Prepare *large* test tubes for your standard curve and your unknowns, the latter of which will be all of your saved LDH fractions.

3. Using the BSA standard, put enough BSA in the tubes to give a range from 0 to 100 μg of protein.

4. In other tubes put in a volume of your fractions. You do not know how concentrated these samples are, so you are just going to have to guess how much to put in, keeping in mind that the maximum volume for this assay is 100 μL. The crude homogenate and anything that is visibly cloudy (i.e., those fractions before the column chromatography) are very concentrated, so you should probably make a 10/1 dilution of those right away and try assaying 10 or 20 μL. It would always be a good idea to assay duplicate tubes. Any duplicates that do not give the same number should be assayed again.

5. Add 3 mL of Bradford reagent to the tubes and vortex immediately. Let the tubes sit for 5 min.

6. While the tubes are sitting, you can zero the spectrophotometer at 595 nm, using tube 1 (zero protein) as the blank.

7. Assay the reactions by pouring roughly 3 mL of solution into a cuvette and reading the absorbance. Pour the solution back out into the large tube as soon as you have recorded the data.

8. Note the absorbance of your 100-μg standard. Any of your LDH samples that have an absorbance higher than that must be reassayed by making up a new tube with less fraction in it until it falls within your standard curve.

9. Calculate the protein concentration of your samples, the specific activity, and the fold purification, making the final purification table.

10. Before you leave, it would be best if you went over a complete set of calculations for one fraction with the teaching staff.

Analysis of Results

Experiment 3a: Protein Determinations

Data

1. Fill in the table below for your standard curve:

Reagent/tube	1	2	3	4	5	6	7	8
BSA volume (μL)	0	10	20	30	40	50	75	100
Protein (μg)								
Absorbance (corrected)								

2. Fill in the following table for your LDH fractions:

Fraction	Dilution Used	Volume Assayed	A_{595}	Protein (μg)	mg/mL
Crude					
20,000 $\times$ g					
65% AS pellet					
Dialyzed 65% pellet					
Pooled IEX fractions					
Dialyzed IEX fractions					
Pooled Cibacron Blue fractions					
Concentrated Cibacron Blue fractions					
Pooled Sephadex fractions					
Concentrated Sephadex fractions					

Analysis of Results

1. Make a proper graph of corrected absorbance vs. μg protein and attach it to this report.

2. Does the trend in your protein concentrations make sense? Why or why not?

3. You can now complete the purification table in Chapter 7. The protein concentration in milligrams per milliliter is divided into the relative activity to give specific activity.

Additional Problem Set

1. The extinction coefficient for NADH is $6220\ M^{-1}\,cm^{-1}$ at 340 nm. Calculate the following:
   a. The absorbance of a $2.2 \times 10^{-5}\ M$ sample in a 1-cm cuvet at 340 nm.
   b. The absorbance of a $2.2 \times 10^{-5}\ M$ sample in a 1-mm cuvet at 340 nm.
   c. The absorbance of a 1 mM sample in a 1-cm cuvet at 340 nm.

2. How would you calculate the extinction coefficient for NADH at 260 nm?

3. Define the terms *transmittance*, *% transmittance*, *absorbance*, and *extinction coefficient*.

4. Calculate the molar extinction coefficient of a biomolecule with a MW of 300 if a 5.0 mM solution in a 1.2 cm cuvette has an absorbance of 0.50 at 340 nm.

5. A compound has an extinction coefficient of $22{,}150\ M^{-1}\,cm^{-1}$. If a solution of this compound has an absorbance of 0.6 in a 1.15 cm cuvette, what is the concentration of the solution?

6. Aliquots of a 0.5 mg/mL standard of BSA were used to construct a standard curve for the Bradford protein assay. The tubes contained the following amounts of the BSA solution: 0, 20, 40, 60, 80, 100 μL. The corresponding absorbencies after adding Bradford reagent were the following: 0, 0.05, 0.09, 0.14, 0.19, and 0.22. If you took 20 μL of an unknown and added 80 μL of water, mixed, took 10 μL of the mixture, added this to Bradford reagent, and saw an absorbance of 0.08, what was the protein concentration of the undiluted unknown?

Webconnections

For a list of Web sites related to the material covered in this chapter, go to **Webconnections** at the *Experiments in Biochemistry* site on the Saunders College Publishing Web page. You can access this page at http://www.saunderscollege.com. **Webconnections** are under *Experiments in Biochemistry* in the Biochemistry portion of the Chemistry page.

References and Further Reading

M. M. Bradford (1976) *Anal. Biochem.* **72**. A rapid and sensitive method for the quantitation of protein.
C. Burgess and K. D. Mielenz, *Advances in Standards and Methodology in Spectrophotometry*, Elsevier Science, Ltd. 1987.
S. Gorog, *Ultraviolet and Visible Spectrophotometry in Pharmaceutical Analysis*, CRC Press, 1995.
R. Cecil Jack, *Basic Biochemical Laboratory Procedures and Computing*, Oxford University Press, 1995.
D. T. Plummer, *An Introduction to Practical Biochemistry*, 2nd ed., McGraw-Hill, 1978.
J. F. Robyt and B. J. White, *Biochemical Techniques*, Waveland Press, 1990.
I. H. Segel, *Biochemical Calculations*, 2nd ed., Wiley Interscience, 1976.
J. Stenesh, *Experimental Biochemistry*, Allyn and Bacon, 1984.
L. Sommer, *Analytical Absorption Spectrophotometry in the Visible and Ultraviolet: The Principles*, Elsevier Science, 1990.

Chapter 4

Enzyme Purification

Topics

Introduction

Enzymes control all of the biochemical reactions in an organism. To understand the metabolism of a cell or organism, one must understand the enzymes responsible. Before studying an enzyme, it must be purified away from contaminants. This chapter begins the study of enzyme purification. The enzyme, lactate dehydrogenase (LDH), will be purified from beef heart tissue via several procedures. This chapter begins with the procedures of homogenization, centrifugation, and ammonium sulfate precipitation.

4.1 Enzymes as Catalysts

Although most of the reactions that occur in your body are possible in thermodynamic terms, they would take place far too slowly to be useful without a catalyst. Enzymes are the biological catalysts that allow metabolism to happen. Without them, the only way for the same reactions to occur would be to raise the temperature to extremely high

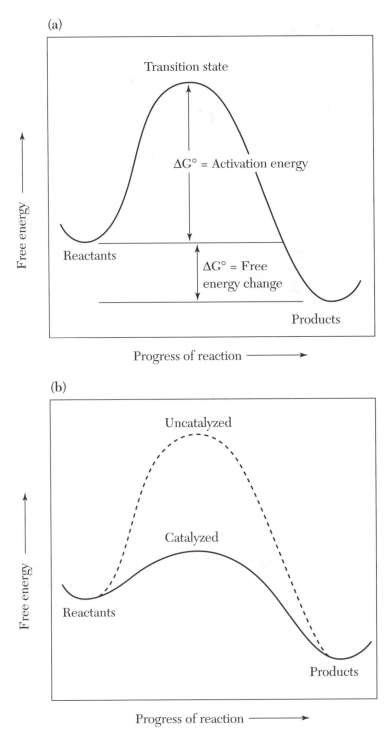

Figure 4.1 Activation profiles for catalyzed and uncatalyzed reactions

levels where life could not exist. Enzymes work by lowering the energy of the transition state between the reactants and the products. Figure 4.1 shows the energy diagram of this effect.

With the exception of some specialized ribonucleic acids (RNAs) that were recently discovered, all enzymes are proteins. These proteins are the most efficient catalysts known. Enzymes can increase the speed of a reaction by up to 10^{20} times that of an uncatalyzed reaction. These proteins are also very specific for the substrates of the reaction. If an enzyme normally reacts with L-alanine, for example, it may not recognize D-alanine at all. Some other enzymes, however, may have several substrates that they recognize and react with.

4.2 Enzyme Purification

Because biological systems are very complex and difficult to study in vivo, enzymes are usually separated from other proteins and metabolites and studied individually. This separation is called **enzyme purification,** and the desired product is a preparation that contains only the enzyme of interest. Usually many techniques are combined, with each one removing more of the unwanted material.

A common way to purify an enzyme is to take a tissue source rich in enzyme and grind it up it in a buffer. This is called **homogenization.** From the moment you do that, you are in a race against time. Inside the cell, the enzyme of interest may be protected from other enzymes whose job it is to degrade proteins. Once you homogenize the tissue, these degradative enzymes, called **proteases,** will begin to degrade all proteins. Biological enzymes function best at around body temperature or 37°C. If we keep the solution cold, say around 4°C, then we can purify the enzyme and keep the proteases from functioning. That is why most enzyme purifications are done in a cold room or on ice the entire time. You hope that as you purify, you will not lose too much of your enzyme, but you will lose most of the contaminating proteins from the homogenate. In the experiment(s) that follow, we will be partially purifying LDH, which we will use for our model purification.

Homogenization of Lactate Dehydrogenase

We will use beef heart as our source of LDH. LDH is an enzyme found in the cell cytosol, which makes it easy to isolate. We start by cutting up some heart muscle into small pieces and trimming off any obvious nonmuscle tissue, such as fat. We then mix the heart pieces with a suitable buffer, usually a phosphate buffer at pH 7.5, in a prescribed ratio. Then we use a blender to grind the heart tissue. This step happens in a 4°C cold room. When we are done, we have a beef heart homogenate suitable for purifying.

Centrifugation

Sometimes particles are separated from others based on differences in density. When particles are spun in a centrifuge, they will move according to their density, the density of the solution, and the force applied. For example, in a crude tissue homogenate, if the force applied

Tip 4.1

⮑ When you are purifying an enzyme, always remember that every second that your sample is not in a cold room or an ice bucket it is susceptible to degradation. The difference between your enzyme preparation and that of the student next to you is usually the care it is given in between the actual purification steps.

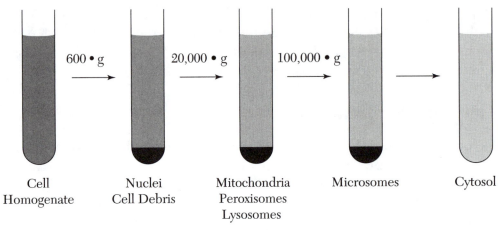

Figure 4.2 Differential centrifugation

is 500 × *g*, unbroken cells and nuclei will precipitate, and the other particles will stay in solution. If you increase the force to 10,000 × *g*, mitochondria, peroxisomes, and lysozomes will spin down. To precipitate the microsomes (fragments of endoplasmic reticulum), you would have to use 100,000 × *g*. Figure 4.2 shows the fundamentals of separating subcellular organelles by centrifugation. Other times, the centrifuge is used to precipitate everything to the bottom.

To control the force applied, it is necessary to know the radius of the centrifuge rotor. The larger the radius, the greater the force at a given speed. Usually we report centrifugations in terms of forces × *g*. You must then figure out how fast to run it in order to get the desired force. Tables exist to convert revolutions per minute for different rotors to *g* forces. Table 4.1 shows a typical conversion chart for rotor speed to force × *g*.

Table 4.1 Rotor Speed versus *g* Force

RPM	JA-21 102	JA-20 108	RPM	JA-21 102	JA-20 108
500	28	30	8,000	7,310	7,740
1,000	114	120	8,500	8,250	8,740
1,500	257	272	9,000	9,250	9,800
2,000	456	483	9,500	10,300	10,900
2,500	714	756	10,000	11,400	12,100
3,000	1,030	1,090	10,500	12,600	13,300
3,500	1,400	1,480	11,000	13,800	14,600
4,000	1,830	1,940	11,500	15,100	16,000
4,500	2,310	2,450	12,000	16,500	17,400
5,000	2,860	3,030	12,500	17,900	18,900
5,500	3,460	3,660	13,000	19,300	20,400
6,000	4,110	4,350	13,500	20,800	22,000
6,500	4,830	5,110	14,000	22,400	23,700
7,000	5,600	5,930	14,500	24,000	25,400
7,500	6,430	6,800	15,000	25,700	27,200

In the purification of LDH, centrifugation will be used three times. The first time will be when you take the crude homogenate and spin it at $20,000 \times g$. This will precipitate out any unbroken cells, nuclei, mitochondria, and most of the peroxisomes and lysozomes. No LDH should be in any of these except the unbroken cells, because LDH is a cytosolic enzyme. Once you have the supernatant, or super, from the $20,000 \times g$ spin, you will add ammonium sulfate and get a precipitate. Centrifugation will be used to spin down this precipitate which should have only contaminating proteins. To the supernatant from that spin, you will add more ammonium sulfate. This will form another precipitate, which should have the LDH. This will then be spun down and collected.

Salting Out

From the $20,000 \times g$ supernatant solution, which contains many contaminating proteins, we must purify the enzyme. A common way is to start with **ammonium sulfate precipitation.** This technique is called **salting out.** Proteins are kept in solution by interactions between their hydrophilic portions and the solvent. Hydrogen bonds form and the proteins are surrounded by the water molecules in the solvent. By adding a very polar compound, such as ammonium sulfate, $(NH_4)_2SO_4$, many of the water molecules will interact with the salt instead of with the proteins. With less water available to stabilize the proteins, the proteins begin to interact with each other and clump together, much as oil clumps together instead of mixing with water.

Certain proteins will precipitate out of solution at low concentrations of ammonium sulfate; others require higher concentrations. We will add a low concentration of ammonium sulfate to our homogenized beef heart, allow the proteins to precipitate, and centrifuge to drive the precipitated proteins to the bottom. To precipitate the LDH, we will add more ammonium sulfate to the supernatant. This will then be centrifuged and the pellet collected.

When you are doing an ammonium sulfate precipitation, it is best to use either a saturated solution or ammonium sulfate crystals that have first been ground into a powder. This allows for quicker mixing without getting locally high concentrations of the salt, which will cause the wrong proteins to precipitate out. The powder must be added slowly with good mixing for the same reasons.

Further Purification Steps

If you wanted to purify LDH to 100% purity, you would have to choose several other techniques. A good combination would be to use several chromatography techniques, such as ion exchange (Chapter 5), gel filtration (Chapter 7), or affinity chromatography (Chapter 6). Each one of these column techniques would provide a sample of increasing purity. In the

Tip 4.2
The best way to get a good ammonium sulfate precipitation is to add the salt after it has been ground into a powder. This way it goes into solution more easily. Also, you want to add it very slowly so you never have a locally higher concentration than you think.

experiment that immediately follows, however, we will only go as far as the ammonium sulfate precipitation.

4.3   Units of Enzyme Activity

Before beginning to purify, you must understand the quantities that we will be measuring. Enzymes catalyze a specific reaction, and what we are interested in is the amount of product formed per unit time. This is called the **activity.**

$$\text{Activity} = \text{Amount of product/Time}$$

Usually an enzyme activity is measured in micromoles of product per minute. This would be one **unit of activity.**

$$\text{1 unit of activity (U)} = 1 \ \mu\text{mol/min}$$

However, we can define a unit to be anything we want, depending on the assay used. If you were studying an enzyme with extremely low activity in the cell, you might tire of writing 0.003 units, so you could define 1 unit to be 1 nmol/min. Then you would have 3 units. Thus, when we measure the activity of an enzyme, we are not actually measuring the number of enzyme molecules; rather we are measuring what those molecules are doing. The activity is very dependent on reaction conditions, such as pH, ionic strength, temperature, and the presence of inhibitors or activators.

   If we want to know how concentrated our enzyme sample is, then we might want to calculate how many units of activity we have in a certain volume. This is called **relative activity** and is usually measured in U/mL.

$$\text{Relative activity} = \text{units/mL} = \text{U/mL}$$

   Therefore, if we use the preceding definition of units, add 0.10 mL of an enzyme preparation, and find that the reaction proceeds at 10 μmol/min, the relative activity of the enzyme preparation is

$$\frac{10 \ \mu\text{mol/min}}{0.1 \ \text{mL}} = \frac{100 \ \mu\text{mol/min}}{\text{mL}} = \frac{100 \ \text{U}}{\text{mL}}$$

   The number of units per milliliter tells us how concentrated the enzyme is, but not how pure it is. What we want is a high activity of the enzyme without a lot of other proteins around. To measure this we use **specific activity.**

$$\text{Specific activity} = \text{U/mg Protein}$$

which is the activity in units divided by the number of milligrams of total protein in the sample. Keep in mind that you could have two fractions of equal specific activity that were quite different. Because specific activity is a ratio, anything that increases the number of units or decreases the number of milligrams of protein increases the specific activity. The number of milligrams of protein is a measure of total protein, active or inactive, enzyme or nonenzyme. The number of units is only the live form of the enzyme we want. In general, the fraction with the highest specific activity is considered the most pure.

◁▭▶ Practice Session 4.1

We have three enzyme fractions. We take 200 μL of each and assay for activity. We also take 0.5 mL of each and assay for protein. The results are as follows:

Fraction	Units	Protein (mg)
1	10	10
2	20	10
3	25	20

Calculate the specific activity of each fraction.

For fraction 1, we have 10 units that came from 200 μL or 0.2 mL, so we have 10U/0.2 mL = 50 U/mL. Furthermore 0.5 mL of my fraction contained 10 mg of protein, so 10 mg/0.5 mL = 20 mg/mL.

$$\textbf{Specific Activity} = \text{U/mg} = \frac{\text{U/mL}}{\text{mg/mL}} = \frac{50}{20} = 2.5 \text{ U/mg}$$

Following the same logic for the other fractions gives 5 U/mg and 3.125 U/mg, respectively.

Two other important quantities are the **% recovery** and the **fold purification.** There is always a starting point in a purification, and as you purify, you compare the fractions to the starting point. The starting point is usually called the crude.

$$\textbf{% Recovery} = \frac{\text{Total units of fraction}}{\text{Total units of crude}} \times 100$$

That is, if we started with 1000 U in the crude and we have 500 U in the fraction, there is a 50% recovery at that point.

$$\textbf{Fold purification} = \frac{\text{Specific activity of fraction}}{\text{Specific activity of crude}}$$

Therefore, if the purified fraction has a specific activity of 2000 U/mg and the crude had one of 100 U/mg, the fold purification is 20.

Usually the fold purification increases while the % recovery decreases during a purification. Sometimes, though, you may see strange things, like a % recovery that is over 100%. Don't panic. This happens for two common reasons. First, you are comparing all of your fractions to a crude sample. Crude samples are difficult to measure because they contain too many contaminating proteins and particulates, so your measurement of the crude may be the least accurate measurement you have. Second, remember that you are measuring an enzyme activity, not the number of enzyme molecules. Clearly, you could not gain molecules of LDH, but you could gain LDH activity if each molecule became more active. This happens during a purification when you remove an inhibitor or put the enzyme into a more favorable buffer or salt solution.

4.4 Calculating Initial Velocity

Whenever we assay an enzyme, we want to measure the initial velocity. This is the velocity or number of units that you calculate at the beginning of the reaction. The rate of reaction is dependent on the concentration of substrates. When it is the enzyme we want to measure, we try to hold the other variables constant, so we use large quantities of substrate. When the enzyme encounters the substrates, it converts them to products. The reaction will eventually slow down due to substrate depletion and an opposing back-reaction from the built-up products. The easiest way to measure the initial velocity would be to graph the absorbance vs. time for the reaction. If we measured the velocity of an enzyme-catalyzed reaction, we might see the data shown in Table 4.2.

If you graphed these data, they would appear as in Figure 4.3. You can see that the reaction rate is linear for the first minute, but then decreases. Many students make the mistake of calculating their enzyme rates by taking the absorbance at 2 min, subtracting the absorbance at time zero, and then dividing by 2. This would give an absorbance change per minute, but it would not be the initial velocity, because absorbencies were included that were not on the linear portion. Doing the calculation that way would give you an absorbance change of 0.32/2 = 0.16 per minute. This number is low. The real initial velocity is 0.2 per minute. Calling 0.16 your initial velocity is akin to admitting to the highway patrolman that you were going 85 mph between Ft. Collins and Cheyenne, but that he shouldn't ticket you because you were planning on only doing 40 mph between Cheyenne and Laramie.

The safest way is to graph the data and take the initial slope of the line. That way always works to give you initial velocity. Another way is to just look at the data on the table. You can see that the absorbance is changing by 0.05 for every 15 s and that this change is constant. From those data, you could conclude that your absorbance change is 0.2/minute. Once you have calculated the initial velocity, you can convert to the other units mentioned in section 4.3, such as relative activity, total activity, % recovery, and so on. To do this, you must keep careful records regarding the amount of sample you put into the cuvette to assay and any dilutions you made previous to that. Without that information, you will never be able to do the calculations.

Table 4.2 Absorbance versus Time

Time (sec)	Absorbance
0	0
15	0.05
30	0.10
45	0.15
60	0.20
75	0.24
90	0.27
105	0.30
120	0.32

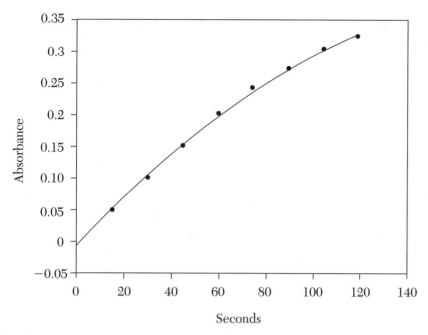

Figure 4.3 Absorbance vs. time for enzyme-catalyzed reactions

⇨ Practice Session 4.2

If you were doing an enzyme assay that generated the data shown in Table 4-2, what would be a good way to record your data?

The data in Table 4.2 are presented in a good way, but the information is incomplete. This might have been an assay of a particular fraction of your purification. A good way to present the data would be

Time (sec)	Absorbance
0	0
15	0.05
30	0.10
45	0.15
60	0.20
75	0.24
90	0.27
105	0.30
120	0.32

Fraction: 40% ammonium sulfate supernatant

Sample volume: 20 μL

Dilution factor: 10/1

Total volume of fraction: 38 mL

Absorbance per minute: 0.2/min

Initial velocity: 0.096 μmol/min

ESSENTIAL INFORMATION

When we study enzymes, we must calculate how much enzyme we have. We do that by calculating the units of enzyme activity. This is an indirect method based on the measurement of the amount of product the enzyme can produce from substrate per unit time. In effect, we are measuring what the enzyme does rather than what it is. To make this measurement, we calculate a change in absorbance vs. time for the reaction where substrate goes to product. We must be sure that we measured the rate at the beginning of the reaction, which is called the initial velocity. The most secure way to do this is to graph absorbance vs. time for each enzyme reaction and take the initial slope as your velocity. This can then be converted to enzyme units.

4.5   Purification Tables

During the course of a purification, we usually make a purification table to help keep track of the success of the process. A purification table tracks all of the quantities listed in Section 4.3 as we proceed from the crude homogenate to the final product. Table 4.3 shows a sample purification table.

Table 4.3 Sample Purification Table

Fraction	Units/mL	Total Units	% Recovery	Protein (mg/mL)
Crude homogenate	45	2000	100	8.5
20,000 × g super	45.5	1989	99	8.0
40% Ammonium sulfate super	48	1824	91	7.0
65% Ammonium sulfate pellet	212	1224	61	12.6
Ion exchange column pooled fractions	186	1300	65	0.8

Fraction	Units/mg	Fold Purification
Crude homogenate	5.3	1
20,000 × g super	5.7	1.1
40 % Ammonium sulfate super	6.9	1.3
65% Ammonium sulfate pellet	16.8	3.2
Ion exchange column pooled fractions	233	44

4.6 Assay for Lactate Dehydrogenase (LDH)

LDH catalyzes the following reaction:

$$\textbf{Pyruvic Acid + NADH} \rightleftharpoons \textbf{NAD}^+ + \textbf{L-Lactic acid}$$

We will measure the rate of reaction by monitoring the appearance of NADH, which absorbs light strongly at 340 nm. By watching the absorbance going up and timing how fast it is going up, we can calculate how much NADH is being formed per time, which we can convert to units of enzyme activity.

We will be setting up a reaction cocktail containing an assay buffer, lactate, and NAD$^+$. This reaction cocktail is clear, so you will have no visible way to know if it is correct or not. The starting absorbance of the cocktail should be around zero compared to water. If it is much higher than that, the cocktail is faulty, probably due to NADH that has already been produced. You must be *very careful* that no pipet that ever touched LDH is allowed to contact your reaction cocktail. Use a clean, new pipet tip every time.

Converting from Absorbance to Units

For the LDH assay, we use the extinction coefficient of NADH to convert from absorbance change per minute to units. Our standard assay has a volume of 3 mL (0.003 L). We report units in micromoles of product per minute. Therefore, the following equation is used to calculate units:

$$\text{Units} = \frac{(\Delta \text{ absorbance}/\Delta \text{ min})}{6220 \text{ M}^{-1} \text{ cm}^{-1} \text{ (cm)}} \times \frac{10^6 \text{ }\mu\text{M}}{\text{M}} \times 0.003 \text{ L} = \mu\text{mol/min}$$

The standard assay will not change during these experiments, so we can combine most of the information into a constant, which turns out to be 0.48. Thus,

$$\text{Units} = \frac{\mu\text{mol}}{\text{min}} = \frac{\Delta \text{ absorbance}}{\Delta \text{ min}} \times 0.48$$

4.7 Why Is This Important?

We are using a well-known enzyme, LDH, to study centrifugation and enzyme purification. The most important thing you will get out of this is the ability to use a preparative centrifuge. Although this is not difficult to use, it is important that it be used correctly. Most of what we know about metabolism came from the combination of enzyme purifications done on the individual enzymes responsible. You cannot study glycolysis or the Krebs cycle as a whole. The individual pieces must be studied.

The next time you are taking part in your favorite sport and you "feel the burn," you will know that lactic acid is being produced. I am sure that you will quickly say to yourself, "My legs are burning because I am utilizing glucose anaerobically and lactic acid is the final product, which makes my legs hurt," instead of just saying, "Ouch." Try it; it works.

You are also learning the early steps and theories behind classical enzyme purification. There is no magic to purifying an enzyme; rather it is the combination of a logical sequence of steps designed to rid your sample of all contaminants. Starting with the simple techniques in this chapter and adding a few more that we will see in subsequent chapters, you will have a reasonable shot at purifying an enzyme to homogeneity.

4.8 Tricks of the Trade

Hints for Getting Good Enzyme Rates

1. Do not vortex enzymes. Invert the cuvet three times after adding the enzyme.

2. Have the cuvet with reaction cocktail warmed to room temperature before adding the enzyme.

3. Avoid using 100 μL of your samples, because they are very crude and there will be a lot of light scattering. Start with 20 μL of sample plus 80 μL of extra water and see if the rate is measurable.

Hints For Being Able to Do the Calculations Later

Whenever you do enzyme purifications, you end up doing many calculations. For each assay you do, you need some essential pieces of information to do those calculations:

1. Record the volume of enzyme that went into the assay vessel. Note that this is not 100 μL each time. If we add 80 μL of water to the assay vessel and then add 20 μL of enzyme sample, the number we record is 20 μL.

2. Record any dilutions you made. This is probably the most confusing part to those who do not do this for a living. Following the preceding example, we would not say we made a dilution just because we put 80 μL of water in first and then the 20 μL of enzyme. This was *not* a 5/1 dilution. We actually assayed 20 μL of undiluted sample. A dilution is when we have to make up a new, separate sample before we even try to put some of it into the assay vessel.

   If I take 100 μL of 20,000 $\times$ g supernatant and add 900 μL of homogenization buffer, we have made a 10/1 dilution. If we then take 20 μL of that dilution and put it into an assay vessel with 80 μL of water and 2.9 mL of reaction cocktail, we have assayed 20 μL of a 10/1 dilution.

3. Record how many total mL of each fraction you have. You eventually will need to know that you had 35 mL of 20,000 $\times$ g supernatant or 40 mL of 65% supernatant, and so on.

Experiment 4

Purification of Lactate Dehydrogenase (short version)

In this experiment, you will partially purify lactate dehydrogenase (LDH) from beef heart using the techniques of homogenization, centrifugation, and ammonium sulfate precipitation. You will also learn how to do the assay for LDH.

Prelab Questions (week 1)

1. What is the basis for the LDH assay?

2. How do we know if our reaction cocktail has been compromised?

3. What must be done just before putting your centrifuge tubes into the centrifuge?

4. Why do you add *ground* ammonium sulfate and why do you add it *slowly*?

Prelab Questions (week 2)

1. Draw the protocol you will use for the protein standard curve. How many milliliters of 0.5 mg/mL BSA will you use in each tube to get the range of 0–50 μg that you want?

2. What size and number of tubes will you use for the various parts of the experiment?

Objectives

1. Explain simple purification schemes for enzymes.

2. Explain how salting out with ammonium sulfate allows biological molecules to be separated.

3. Utilize centrifugation for separation of biological preparations.

4. Dilute enzyme preparations and assay for LDH activity.

5. Construct standard curves and determine protein concentrations of enzyme fractions by the Bradford method.

6. Determine the specific activity of enzyme fractions and construct a purification table for the purification of LDH.

Experimental Procedures

Materials

Beef heart crude homogenate

Assay buffer, 0.15 M CAPS, pH 10.0

NAD^+, 6 mM

Lactic acid, 150 mM

Ammonium sulfate

BSA standard solution, 0.5 mg/mL

Bradford reagent for protein determination

Methods

Lab Period 1

All procedures will be carried out on ice.

1. Acquire 35 mL of ground beef heart in buffer and put it in a beaker on ice as quickly as possible. *Save 1 mL for assays and protein determinations before going on to step 2.*

2. Put the homogenate into centrifuge tubes or bottles as directed in lecture and balance with another student's bottle or with a bottle of water. *The balancing is of the utmost importance!*

3. Centrifuge at 20,000 × g for 20 min. The centrifuges will be set at 4°C.

4. While the centrifugation is going, assay the crude homogenate for enzyme activity. The reaction cocktail for the assay will be the following:

    Assay buffer, 1.9 mL

    Lactate, 0.5 mL

    NAD^+, 0.5 mL

    This is the amount for one assay, so make up multiples of that depending on the number of assays you plan to do. Zero the spectrophotometer with water at 340 nm or as close to 340 as your machine will allow. Measure the absorbance of the reaction cocktail to be sure it is close to zero.

5. To the 2.9 mL of cocktail, add extra assay buffer and then LDH from your crude homogenate to bring the volume up to 3 mL. Start with 10 μL of sample. Invert three times and measure the absorbance changes immediately.

6. Use whatever dilutions are necessary to get an accurate change per minute. Dilute with assay buffer (not reaction cocktail or water). Be sure to record your dilutions.

7. Freeze the leftover crude homogenate sample for next week. Also save your dilutions.

8. When the centrifugation is over, discard the pellet and save the supernatant.

9. Measure the volume of the supernatant. *Save 1 mL for enzyme assay* and *slowly* add 0.230 g of ground (powder, not crystal) ammonium sulfate for every milliliter of supernatant. Stir the solution constantly while you add the salt. This amount of ammonium sulfate will bring the percent saturation to 40. This is called taking a 40% cut.

10. Let the solution stand (on ice as always) for 10 min while the proteins precipitate.

11. Balance the tubes and centrifuge at $15,000 \times g$ for 15 min.

12. Discard the pellet. Save and measure the supernatant. Take 1 mL of the supernatant and assay for LDH activity as before.

13. To the rest of the supernatant, add 0.166 g ground ammonium sulfate for every mL of supernatant. Add the salt slowly with constant stirring as before. This amount of ammonium sulfate will bring the final concentration up to 65% saturated.

14. Let the solution stand as before. While this is going on, you can assay the 40% supernatant that you saved.

15. Spin the 65% solution as before. Save the supernatant and pellet.

16. Resuspend the pellet in a small quantity of homogenization buffer (2–5 mL).

17. Measure the volume of the resuspended 65% pellet and assay for LDH activity. Assay the 65% supernatant too.

18. Save all of your samples in microfuge tubes (crude homogenate, $20,000 \times g$ supernatant, 40% supernatant, 65% pellet, and 65% supernatant). These will be used next week for protein determinations.

19. At this point in the experiment, you can calculate the activity (units in the reaction cuvette), relative activity (units per mL of sample), and total activity (total units of sample) for each sample.

Lab Period 2

1. Make a protein standard curve for the Bradford assay (see Chapter 3) using the 0.5 mg/mL BSA solution. You want 0–50 μg BSA in the tubes. At the same time, set up protein assays for your five fractions from last time. You will probably need small quantities, such as 10 μL, but you will have to play around with that.

2. Add 3 mL Bradford reagent, vortex immediately, and let stand for 10 min. Read the absorbencies at 595 nm.

3. If your sample absorbencies are off of your standard curve, remake the tubes using less sample or a diluted sample.

4. At this point, you can calculate the specific activity (units/mg protein) of the fractions, percent recovery ([total units fraction/total units crude homogenate] $\times$ 100), and the fold purification (specific activity fraction/specific activity crude homogenate).

Analysis of Results

Experiment 4: Purification of Lactate Dehydrogenase (short version)

Data

Period 1

1. Provide all of the information on your purification steps:

    Milliliters of crude homogenate used _____

    Milliliters of 20,000 × g super _____

    Grams of $(NH_4)_2SO_4$ used for 40% cut _____

    Milliliters of 40% super _____

    Grams of $(NH_4)_2SO_4$ used for 65% cut _____

    Milliliters of 65% super _____

    Milliliters of 65% pellet after resuspending _____

2. Provide all of the information on your enzyme assays:

Fraction Isolated	Sample Assayed (μL)	Dilution Made (if any)	Absorbance Change per minute
Crude homogenate			
20,000 $\times$ g super			
40% $(NH_4)_2SO_4$ super			
65% $(NH_4)_2SO_4$ pellet			
65% $(NH_4)_2SO_4$ super			

Period 2

1. Provide all of the information on your samples and the Bradford assay.

Standard Curve

Volume BSA	Absorbance

Bradford Assay on LDH samples

LDH Fraction	Quantity Assayed (μL)	Dilution Used	Absorbance
Crude homogenate			
20,000 $\times$ g super			
40% AS super			
65% AS pellet			
65% AS super			

Calculations

1. For each of your fractions, calculate the activity in the reaction vessel (units in μmol/min) and the relative activity of the sample (units/mL of fraction). The extinction coefficient is 6220 M^{-1} cm^{-1}. The cuvettes are 1 cm in diameter.

$$\text{Units} = \frac{\Delta \text{ A}/\Delta \text{ min}}{6220 \text{ M}^{-1} \text{ cm}^{-1} \text{ (1 cm)}} \times \frac{10^6/\mu\text{M}}{\text{M}} \times 3 \times 10^{-3} \text{ L}$$

U/mL = (units from above/volume of fraction assayed) $\times$ dilution used, if any.

2. Calculate the total number of units for each fraction by multiplying the relative activity by the total volume of the fraction.

3. Calculate the % recovery of the purified fractions compared to the crude homogenate.

4. Calculate the protein concentration of the fractions in milligrams per milliliter using the Bradford assay standard curve. Attach your graph of absorbance at 595 nm vs. μg protein.

5. Calculate the specific activity of each fraction by dividing the relative activity by the quantity of protein in mg/mL.

6. Calculate the fold purification by dividing the specific activities of the purified fractions by that of the crude homogenate.

All of these calculations may be summarized in the table at the top of the next page.

Fraction	Units	Units/mL	Total Units	% Recovery	Protein (mg/mL)	Specific Activity	Fold Purification
Crude homogenate				100			1
20,000 $\times$ g super							
40% AS super							
65% AS pellet							
65% AS super							

Questions

1. Five μL of a sample that was diluted 6/1 was assayed for LDH activity. The activity in the reaction vessel, which had a volume of 3 mL, was 0.3 U. What was the $\Delta A/\Delta$min observed? What is the relative activity of the original sample?

2. Ten μL of a 5/1 dilution of the original "undiluted" sample from question 1 was used to measure protein concentration. With the use of a standard curve this was found to be 140 μg. What is the specific activity of the original sample?

Experiment 4a

Purification of Lactate Dehydrogenase (comprehensive version)

In this experiment, you will begin a comprehensive experiment about protein purification. Lactate dehydrogenase (LDH) will be isolated from beef heart by homogenization and centrifugation. It will be further purified by ammonium sulfate precipitation and dialysis.

Prelab Questions (week 1)

1. What is the basis for the LDH assay?

2. How do we know if our reaction cocktail has been compromised?

3. What must be done just before putting your centrifuge tubes into the centrifuge?

4. Why do you add *ground* ammonium sulfate and why do you add it *slowly*?

Prelab Questions (week 2)

1. Draw the protocol you will use for the protein standard curve. How many milliliters of 0.5 mg/mL BSA will you use in each tube to get the range of 0–50 μg that you want?

2. What size and number of tubes will you use for the various parts of the experiment?

Objectives

1. Explain simple purification schemes for enzymes.

2. Explain how salting out with ammonium sulfate allows biological molecules to be separated.

3. Use centrifugation for separation of biological preparations.

4. Dilute enzyme preparations and assay for LDH activity.

5. Begin a purification table for the purification of LDH.

Experimental Procedures

Materials

Beef heart

Assay buffer, 0.15 M CAPS, pH 10.0

NAD^+, 6 mM

Lactic acid, 150 mM

Ammonium sulfate

Homogenization buffer, 0.05 M sodium phosphate, pH 7.0

Q-Sepharose® buffer, 0.03 M bicine, pH 8.5

Methods

All of the following should be done in a cold room or on ice unless it just isn't possible. Each experiment should start with about 100 mL of crude homogenate. Some of the steps may be done in groups to account for the number of rotor spaces and the size of centrifuge bottles being used.

At each step of a purification, it is important to record the volume of the sample you are working on. Also, save 0.5 mL for enzyme assays and future protein determinations. After the enzyme assays, freeze the leftover samples for experiment #3a.

Preparation of LDH from Crude Tissue

1. Cut up the beef heart into fine pieces, being sure to avoid the obvious fatty deposits and connective tissue.

2. Combine 25 g of beef heart with 75 mL of cold 0.05 M sodium phosphate, pH 7.0, and homogenize at high speed in the blender for 2 min at 4°C. *Save 0.5 mL for enzyme and protein assays.*

3. Spin the homogenate at $20,000 \times g$ for 15 min at 4°C. *Save 0.5 mL of the supernatant for assays.* This will be known as your $20,000 \times g$ super. The pellet can be safely discarded.

4. Slowly add *ground* ammonium sulfate to the supernatant so that it becomes a 40% saturated solution. This requires 0.242 g of ammonium sulfate for every mL of supernatant you have. Make sure the ammonium sulfate is mixed in slowly and completely and let the sample sit on ice for 15 min.

5. Centrifuge at $15,000 \times g$ for 15 min at 4°C. *Save 0.5 mL of the supernatant* as the *40% super.* Discard the pellet.

6. To the rest of the supernatant, add ammonium sulfate until the final concentration makes a 65% saturated solution. This requires 0.166 g ammonium sulfate/mL. Let this stand for 15 min as before. Spin at $15,000 \times g$ as before.

7. *Save 0.5 mL of the supernatant* as your *65% super.* Note that this fraction should not have significant activity, but save the rest of the supernatant just to be sure.

8. To the pellet from the 65% cut, add 5–10 mL of 0.03 M bicine, pH 8.5, and dissolve the pellet well. *Save 0.2 mL as your 65% pellet.*

9. Place the remainder of the 65% redissolved pellet into a dialysis bag and suspend in a bucket of 0.03 M bicine (Q-Sepharose buffer) in a 4°C cold room.

Assay for LDH

Each of the fractions listed (shown in italics) needs to be assayed today. Enzymes are notoriously unstable in crude form, and we need to find out how much is there before it gets chopped up by proteases.

LDH will be assayed by monitoring the formation of NADH, which absorbs at 340 nm. Each assay will have a reaction volume of 3 mL and will contain the following:

1.9 mL of CAPS buffer, 0.14 M, pH 10

0.5 mL of NAD$^+$, 6 mM

0.5 mL of lactate, 0.15 M

The reaction is initiated by adding 100 μL of enzyme sample plus water. For your cruder samples, you will probably need to start with 10 μL of sample and 90 μL of water. Even 10 μL may be too much.

Whenever possible, use reaction cocktails. If you figure you are going to do 10 assays, then mix up 10 assays worth of ingredients into a flask:

10 × 1.9 mL CAPS = 19 mL

10 × 0.5 mL Lactate = 5 mL

10 × 0.5 mL NAD$^+$ = 5 mL

Once that is made up, just pipet 2.9 mL of it each time into the assay vessel. Then add the 100 μL of enzyme sample. This 100 μL may be all enzyme sample or may be some water and some enzyme sample, always totaling 100 μL. Add the extra water to the cocktail first, then add the enzyme to initiate the reaction. Invert the cocktail three times and put it in the spectrophotometer.

Recovery Tables

You will eventually make a recovery table for your LDH purification. For each fraction that you have (20,000 × g super, 40% super, etc.) calculate the units, units/per milliliter, total units in the fraction, % recovery, specific activity, and fold purification. It might help you to also have columns for sample volume assayed, dilution factors, and $\Delta A/\Delta$ min.

For example, if you had 20 mL of 20,000 × g super, put 20 μL of it along with 80 μL buffer into your reaction vessel, and saw a ΔA/min of 0.31, your calculations would look like this:

$$\text{Units} = (0.31/6220 \text{ M}^{-1}) \times 10^6 \ \mu\text{M/M} \times 0.003 \text{ L} = 0.15 \text{ U}$$

$$\text{Units/mL} = 0.15 \text{ U}/0.02 \text{ mL} = 7.5 \text{ U/mL}$$

If you had made a dilution of 10/1 before you put the 20 μL into the reaction vessel, this number would be multiplied by 10.

$$\textbf{Total units} = \textbf{7.50 U/mL} \times \textbf{20 mL} = \textbf{150 U}$$

$$\textbf{\% Recovery} = \textbf{150 U/150 U} = \textbf{100\%}$$

By definition, the % recovery is 100% for whatever you started with, which is the 20,000 $\times$ g super in this case. For the other samples, you divide the total number of units of the sample by the total number of units of the 20,000 $\times$ g super and multiply by 100 to get % recovery.

After doing the protein determinations on the last day, you can calculate the rest. If you put 20 μL of a 10/1 dilution of 20,000 $\times$ g super into the Bradford protein assay, and your standard curve told you there were 3 μg of protein there, then your protein concentration would be as follows:

$$\textbf{Protein concentration} = \textbf{mg/mL} = \textbf{(3 }\mu\textbf{g/20 }\mu\textbf{L)} \times \textbf{10} = \textbf{1.5 mg/mL}$$

Specific activity is the activity per milligram of protein, which is most easily arrived at by taking the relative activity and dividing by the protein concentration:

$$\textbf{Specific activity} = \textbf{U/mg} = \frac{\textbf{U/mL}}{\textbf{mg/mL}} = \frac{\textbf{7.5 U/mL}}{\textbf{1.5 mg/mL}} = \textbf{5 U/mg}$$

Fold purification is the ratio of the specific activity of a given fraction divided by the specific activity of the 20,000 $\times$ g super.

Analysis of Results

Experiment 4a: Purification of Lactate Dehydrogenase (comprehensive version)

Data

1. Provide all of the information on your purification steps:

    Milliliters of crude homogenate used _____

    Milliliters of 20,000 $\times$ g super _____

    Grams of $(NH_4)_2SO_4$ used for 40% cut _____

    Milliliters of 40% super _____

    Grams of $(NH_4)_2SO_4$ used for 65% cut _____

    Milliliters of 65% super _____

    Milliliters of 65% pellet after resuspending _____

2. Provide all of the information on your enzyme assays:

Fraction Isolated	Quantity of Sample Assayed (μL)	Dilution Made (if any)	Absorbance Change per Minute
Crude homogenate			
20,000 × g super			
40% $(NH_4)_2SO_4$ super			
65% $(NH_4)_2SO_4$ pellet			
65% $(NH_4)_2SO_4$ super			

Calculations

1. For each of your fractions, calculate the activity in the reaction vessel (units in μmol/min) and the relative activity of the sample (units/mL of fraction). The extinction coefficient is 6220 M^{-1} cm^{-1}. The cuvettes are 1 cm in diameter.

$$\text{Units} = \frac{\Delta \text{ A}/\Delta \text{ min}}{6220 \text{ M}^{-1} \text{ cm}^{-1} \text{ (1 cm)}} \times 10^6 \ \mu\text{M/M} \times 3 \times 10^{-3} \text{ L}$$

U/mL = (units from above/volume of fraction assayed) × dilution used, if any.

2. Calculate the total number of units for each fraction by multiplying the relative activity by the total volume of the fraction.

3. Calculate the % recovery of the purified fractions compared to the crude homogenate.

All of these calculations may be summarized in the following table:

Fraction	Units	Units/mL	Total Units	% Recovery
Crude homogenate				
20,000 × g super				
40% AS super				
65% AS Pellet				
65% AS supernatant				

Additional Problem Set

1. If bovine lactate dehydrogenase is known to have a temperature optimum around 37°C, why is it important to purify it at 4°C?

2. Explain why you should not lose any LDH activity when you centrifuge a crude sample at 20,000 × g.

3. Explain why two different percentage saturation levels of ammonium sulfate are used in these experiments.

4. Explain the physical interactions of the molecules that lead to proteins falling out of solution in high salt. What is the driving force of protein precipitation?

5. Explain why it is important to add ammonium sulfate slowly.

6. An enzyme-catalyzed reaction produces a product that has a maximum absorbance at 412 nm. The extinction coefficient is 4000 $M^{-1}\,cm^{-1}$. A 10/1 dilution of the enzyme is made, and 20 μL of the dilution are put into a cuvette with 80 μL of water and 1.9 mL of reaction cocktail. The absorbance change per minute is 0.05.
    a. How many units were in the cuvette if 1 unit = 1 $\mu mol/min$?
    b. How many units were in the cuvette if 1 unit = 1 nmol/min?
    c. What is the relative activity of the undiluted enzyme (1 unit = 1 $\mu mol/min$)?

7. An enzyme sample contains 24 mg protein/mL. Twenty μL of this sample in a standard incubation volume of 0.1 mL catalyzed the incorporation of glucose into glycogen at a rate of 1.6 nmol/min. Calculate the velocity of the reaction in terms of
    a. $\mu mol/min$
    b. $\mu mol/L/min$
    c. $\mu mol/mg$ protein/min
    d. units/mL
    e. units/mg protein

8. Fifty mL of the sample in Problem 7 were fractionated by ammonium sulfate precipitation. The fraction precipitating between 30 and 50% saturation was redissolved in a total volume of 10 mL and dialyzed. The solution after dialysis had 12 mL and contained 30 mg protein/mL. Twenty μL of the purified fraction catalyzed the reaction rate of 5.9 nmol/min under the standard assay conditions. Calculate
    a. The recovery of enzyme after the ammonium sulfate step.
    b. The fold purification after the ammonium sulfate step.

Webconnections

For a list of Web sites related to the material covered in this chapter, go to **Webconnections** at the *Experiments in Biochemistry* site on the Saunders College Publishing Web page. You can access this page at http://www.saunderscollege.com. **Webconnections** are under *Experiments in Biochemistry* in the Biochemistry portion of the Chemistry page.

References and Further Reading

R. F. Boyer, *Modern Experimental Biochemistry,* Addison-Wesley, 1993.

P. D. Boyer, The Enzymes, 3rd ed. vol. 1–13, Academic Press, 1970–1976.

A. R. Clarke, H. M. Wilks, D. A. Barstow, T. Atkinson, W. N. Chia, and J. J. Holbrook (1988), *Biochemistry* **27.** An Investigation into the Contribution Made by the Carboxylate Group of an Active Site Histidine–aspartate Couple to Binding and Catalysis in Lactate Dehydrogenase.

R. A. Copeland, "A Practical Introduction to Structure Mechanisms and Data Analysis," in *Enzymes,* VCH Publishers, 1996.

U. M. Grau, W. E. Trommer, and M. G. Rossman (1981), *J. of Mol. Biol.* **151.** Structure of the Active Ternary Complex of Pig Heart Lactate Dehydrogenase.

J. J. Holbrook, A. Liljas, S. J. Steindel, and M. G. Rossmann, "Lactate Dehydrogenase," in *The Enzymes,* 3rd ed., vol. 11, 1975. Academic Press, Boyer, P. D., ed.

G. Kopperschlager and J. Kirchberger, (1996), *J. Chromatogr.,* **684,** Methods for the Separation of Lactate Dehydrogenases and Clinical Significance of the Enzyme.

R. K. Scopes, *Protein Purification: Principles and Practice,* Springer-Verlag Inc., 1994.

J. J. Sedmack and S. F. Grossberg (1977), *Anal. Biochem.* **79.** Protein Determinations.

H. Taguchi and T. Ohta (1991), *J. Biol. Chem.* **266.** D-lactate Dehydrogenase is a Member of D-isomer-Specific 2-Hydroxyacid Dehydrogenase Family.

H. Taguchi and T. Ohta (1994), *J. Biol. Chem.* **268.** Histidine 296 is Essential for the Catalysis in *Lactobacillus plantarum* D-lactate Dehydrogenase.

Chapter 5

Ion Exchange Chromatography

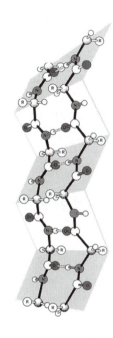

Topics

Introduction

This chapter studies the charged nature of amino acids, peptides, and proteins. Amino acids have weak acid and weak base groups that give them a net positive, negative, or neutral charge depending on the pH environment they are in. This difference in charge can be exploited with the separation technique called **ion exchange chromatography.**

5.1   Amino Acids as Weak Acids and Bases

Amino acids are both weak acids and weak bases because they contain both an amino group and a carboxyl group. When dissolved in water, they exist predominantly in their isoelectric form (no net charge):

$$H_3N^+ - CH - COO^-$$
$$|$$
$$R$$

This form is also called a zwitterion. If you add an acid, it accepts a proton, thereby acting as a base:

$$H_3N^+—CH—COO^- + H^+ \longrightarrow H_3N^+—CH—COOH$$

with R groups below each structure,

base   acid

On the other hand, if you add a base, it acts like an acid and donates a proton:

$$H_3N^+—CH—COO^- + OH^- \longrightarrow H_2N—CH—COO^-$$

with R groups below each structure,

acid   base

If the R group is acidic (Glu, Asp), then it too has a carboxyl group that can donate a proton. If the R group is basic (Lys, Arg, His), then it has an amino group that can accept a proton. An amino acid is like any other weak acid or base except that there can be up to three functional groups to consider. The good news is that you really only consider one of them at a time because at any given pH, usually only one of them is important.

Armed with a table of pK_as and the Henderson–Hasselbalch equation you can work amino acid buffer problems just the same as in Experiment 2.

◁▶ Practice Session 5.1

What forms of glutamic acid will be present at pH 4.3? The pK_as are 2.2, 4.3, and 9.7.
Glutamic acid has the form below at very low pH:

$$\overset{+}{H_3N}—CH—COOH$$
$$|$$
$$CH_2$$
$$|$$
$$CH_2$$
$$|$$
$$COOH$$

You really need only consider the side group carboxyl, but to convince you that this is true, let's determine the % of the functional groups in the A^- and HA forms:

α-COOH pK_a = 2.2

pH = pK_a + log base/acid

4.3 = 2.2 + log base/acid

2.1 = log base/acid

antilog 2.1 = base/acid = 126

In other words, the ratio of the COO^- form to the COOH form is 126 to 1. In essence, all of the α-COOH is dissociated.

α-amino pK_a = 9.7

4.3 = 9.7 + log base/acid

−5.4 = log base/acid

antilog −5.4 = base/acid = 3.9×10^{-6} or in essence, all of it is in the acid form, NH_3^+.

Table 5.1 Properties of the Common Amino Acids

Amino Acid	3-Letter Code	1-Letter Code	pK$_a$ α-COOH	pK$_a$ α-NH$_3^+$	pK$_a$ Side Chain
Alanine	Ala	A	2.34	9.69	
Arginine	Arg	R	2.34	9.69	12.48
Asparagine	Asn	N	2.02	9.82	
Aspartic acid	Asp	D	2.09	9.82	3.86
Cysteine	Cys	C	1.71	10.78	8.33
Glutamine	Gln	Q	2.17	9.13	
Glutamic acid	Glu	E	2.19	9.67	4.25
Glycine	Gly	G	2.34	9.60	
Histidine	His	H	1.82	9.17	6.0
Isoleucine	Ile	I	2.36	9.68	
Leucine	Leu	L	2.36	9.68	
Lysine	Lys	K	2.18	8.95	10.53
Methionine	Met	M	2.28	9.21	
Phenylalanine	Phe	F	1.83	9.13	
Proline	Pro	P	1.99	10.60	
Serine	Ser	S	2.21	9.15	
Threonine	Thr	T	2.63	10.43	
Tryptophan	Trp	W	2.38	9.39	
Tyrosine	Tyr	Y	2.20	9.11	10.07
Valine	Val	V	2.32	9.62	

We didn't really need to calculate the ratio of the base/acid for these two groups because the pH was so far away from the pK$_a$s of the groups. A good rule of thumb is that you can disregard groups if the pH is about 2 units away from the pK$_a$. "Disregard" doesn't mean that the group is not important, rather that it is in one form or the other. With a difference of 2 units, only 1% of the functional group is in the lesser form. Table 5.1 gives the pK$_a$s for the functional groups of the 20 common amino acids. Figure 5.1 gives the structures of the amino acids.

γ-COOH pK$_a$ = 4.3

You should know immediately that at pH 4.3, half of the γ-COOH groups will be dissociated because the pH = pK$_a$. Therefore, at pH 4.3, you have the following forms of glutamic acid:

$$\underset{\overset{|}{COO^-}}{\overset{\overset{NH_3^+}{|}}{CH}}-CH_2-CH_2-COOH \rightleftharpoons \underset{\overset{|}{COO^-}}{\overset{\overset{NH_3^+}{|}}{CH}}-CH_2-CH_2-COO^- + H^+$$

5.2 Isoelectric Point

Another important number to know for an amino acid is its isoelectric point. This takes into account the contribution of the two or three functional groups and tells you the pH at which an amino acid has no net charge. A pH higher than the pI means that there is more hydroxide ion around. More hydroxide ion will pull off more hydrogen ions from the amino acid, leaving you with a negatively charged molecule. Conversely, a pH lower than the pI

(Text continues on page 116.)

(a) Nonpolar side chains

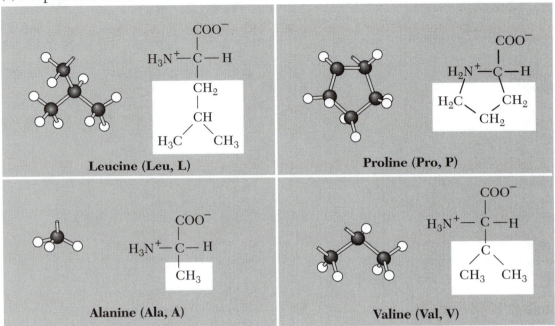

Leucine (Leu, L)

Proline (Pro, P)

Alanine (Ala, A)

Valine (Val, V)

(b) Polar, uncharged side chains

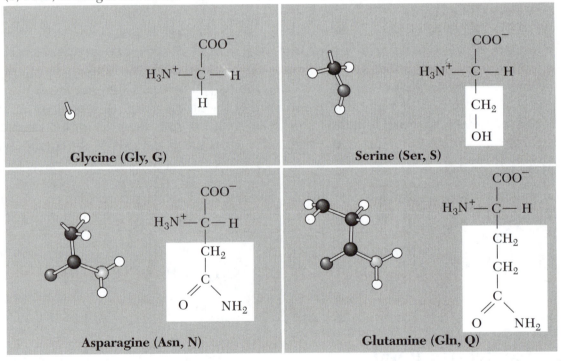

Glycine (Gly, G)

Serine (Ser, S)

Asparagine (Asn, N)

Glutamine (Gln, Q)

Figure 5.1   Structures of the common amino acids

(c) Acidic side chains

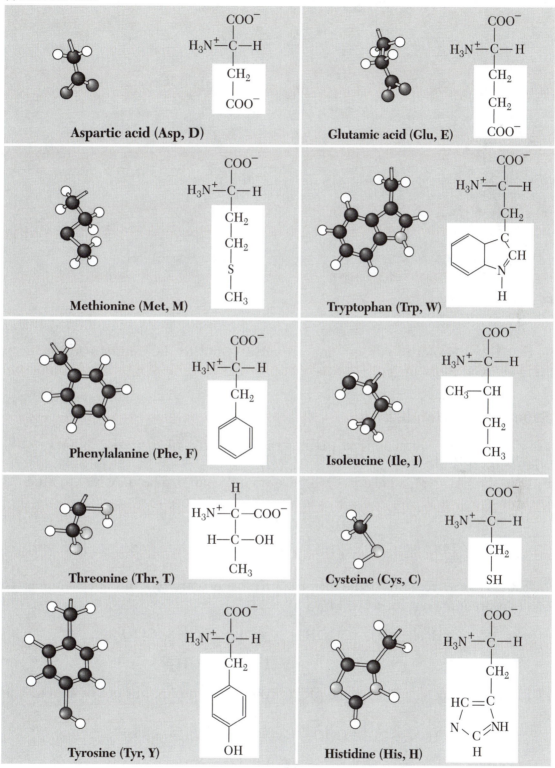

Aspartic acid (Asp, D)

Glutamic acid (Glu, E)

Methionine (Met, M)

Tryptophan (Trp, W)

Phenylalanine (Phe, F)

Isoleucine (Ile, I)

Threonine (Thr, T)

Cysteine (Cys, C)

Tyrosine (Tyr, Y)

Histidine (His, H)

Figure 5.1 Continued

(d) Basic side chains

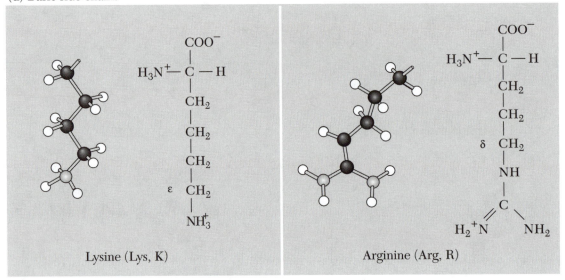

Lysine (Lys, K)   Arginine (Arg, R)

Figure 5.1 Continued

means that there is more hydrogen ion around to bind to the amino acid, giving you a positively charged molecule.

🔊 **Practice Session 5.2**

What is the ionic form of alanine at pH 2, 6, and 10 if the pI is 6, pK_a COOH = 2.3, and pK_a NH_3 = 9.7?

At pH 6, the pH = pI, so alanine will have no *net* charge. The only way to attain this is by the form

$$H_3\overset{+}{N}-CH-COO^-$$
$$|$$
$$CH_3$$

Now, at pH 2 we are more acidic than the pK_a for the carboxyl, but not by much. Therefore the two predominant forms will be

$$H_3\overset{+}{N}-CH-COOH \qquad H_3\overset{+}{N}-CH-COO^-$$
$$| \qquad\qquad\qquad\qquad |$$
$$CH_3 \qquad\quad and \qquad\quad CH_3$$

At pH 10, we are more basic than the pK_a for the amino group, but not by much. The two species will be:

$$H_3\overset{+}{N}-CH-COO^- \qquad H_2N-CH-COO^-$$
$$| \qquad\qquad\qquad\qquad |$$
$$CH_3 \qquad\quad and \qquad\quad CH_3$$

Determining the Isoelectric Point

There is a simple way to determine the isoelectric point of an amino acid. The equation we see written is pI = $(pK_a1 + pK_a2)/2$, but this does not tell us which pK_as to use. Since several amino acids have 3 pK_as, you might accidentally average the wrong ones if you do not have a consistent procedure for doing this. Let's take an amino acid, such as lysine. The pK_as are the following: 2.2 (carboxyl), 9.0 (α-amino), and 10.5 (side-chain amino). Start by drawing the amino acid with all of the hydrogens in place:

$$H_3N^+—CH—COOH$$
$$|$$
$$(CH_2)_4$$
$$|$$
$$NH_3^+$$

Now, remove the hydrogen from the group with the lowest pK_a, because that is where the first hydrogen would come from. Now you have a form that looks like this:

$$H_3N^+—CH—COO^-$$
$$|$$
$$(CH_2)_4$$
$$|$$
$$NH_{3+}$$

Does this form have zero net charge? The answer is no. Therefore, we remove another hydrogen from the group with the next pK_a. Now the molecule looks like this:

$$H_2N—CH—COO^-$$
$$|$$
$$(CH_2)_4$$
$$|$$
$$NH_{3+}$$

Does this form have zero net charge? The answer is yes. To get the pI, average the pK_a of the last group you pulled the hydrogen from (pK_a 9.0) with the next one (pK_a 10.5). This gives you a pI of 9.8.

⬤▷ Practice Session 5.3

What would be the pI for a mythical amino acid that had two carboxylic acid groups (pK_as 2 and 4) and 2 amino groups (pK_as 8 and 10)?

Without even trying to draw such a molecule, we can still calculate the pI. If we start with the molecule with all of the hydrogens in place, then the molecule will have a net charge of +2. If we pull off the hydrogen from the most acidic group (the pK_a 2), we will have one COO^- group, so the net charge will be +1. If we pull off the next hydrogen (the pK_a 4), we will have another COO^- group, so the net charge will be zero. At that point, we average the pK_a 4, which was the pK_a for the last group we removed a hydrogen from, with the pK_a 8, which is the next pK_a in line. The pI is therefore (8 + 4)/2, which equals 6.

This technique for determining pI also works for peptides and proteins. You just have to remember that the majority of the functional groups are tied up in the peptide backbone. Therefore they do not gain or lose hydrogens and are not considered. If you have a peptide with 10 amino acids, there will be one α-carboxyl at the C terminus and one α-amino at the N terminus. The only other groups that count are the side chain acidic or basic groups.

⬤▷ Practice Session 5.4

Calculate the pI for the peptide shown below:

Phe-Lys-Glu-Asp-Lys-Ser-Ala

The first thing to do is redraw this peptide showing clearly what acid or base groups are present. There is the α-amino group on the Phe, the side-chain amino group on the Lys, the side-chain carboxyl on the Glu, the side-chain carboxyl on the Asp, another side-chain amino on the next Lys, nothing on the Ser, and the α-carboxyl group on the Ala:

$$H_3N^+—Phe—Lys—Glu——Asp——Lys—Ser—Ala—COOH$$
$$\qquad\qquad\quad\ \ |\qquad\ \ |\qquad\ \ |\qquad\ \ |$$
$$\qquad\qquad\quad NH_3^+\ \ COOH\ \ COOH\ \ NH_3^+$$

The next step is to figure out the pK_as for the various groups shown. This information is in the table. From left to right they are 9.13, 10.5, 4.3, 3.9, 10.5, and 2.3.

The next step is to remove hydrogens from the various groups starting with the lowest pK_a until you have the form with no net charge. From the figure, we can see that there are three positive charges. Thus we have to remove hydrogens from three groups that will give negative charges in order to have the isoelectric form. When you have done that, the peptide will look like this:

$$H_3N^+—Phe—Lys—Glu——Asp——Lys—Ser—Ala—COO^-$$
$$\qquad\qquad\quad\ \ |\qquad\ \ |\qquad\ \ |\qquad\ \ |$$
$$\qquad\qquad\quad NH_3^+\ \ COO^-\ \ COO^-\ \ NH_3^+$$

Next, you average the pK_a from the last group you removed a hydrogen from (the one with the highest pK_a of those you removed a hydrogen from) with the pK_a of the next group in line (the one with the lowest pK_a of those you did not remove a hydrogen from).

$$pI = (4.3 + 9.0)/2 = \mathbf{6.65}$$

There are many ways of calculating pIs, but no student who learns this method has ever made a mistake!

5.3 Ion Exchange Chromatography

One of the most effective and most used methods for separating charged compounds is ion exchange chromatography, often abbreviated IEX. An ion exchange column is a column full of a resin that contains charged groups on the surface. We will use a type of ion exchange resin called Dowex^{TM} 50, which has a sulfonic acid group as in Figure 5.2.

Originally, the resin is in the acid washed or H^+ form. Then it is reacted with NaOH to strip off the dissociable hydrogens giving the sodium form. It is this sodium form that is the ion exchanger because the sodium ions can be **exchanged** for other positively charged molecules. That is, a positively charged molecule, such as an amino acid below its pI, could bind to one of the sulfonyl groups displacing a sodium ion. Because it is a cation that binds and exchanges, this type of column is also called a **cation exchange column.**

By running a mixture of amino acids over an ion exchange column, you can separate the amino acids. Figure 5.3 demonstrates this principle. Let's assume we have the Dowex column described, already in the sodium form. We then take a mixture of three amino acids (histidine, serine, and aspartic acid) and run it through the column. Everything is in a pH 3.25 buffer. At that pH, the aspartic acid has two forms: 80% is in a form with no net charge, and 20% is in a form with a net negative charge. This means that there is no reason for the aspartic acid to bind to the ion exchanger and it elutes quickly. The serine is electrically

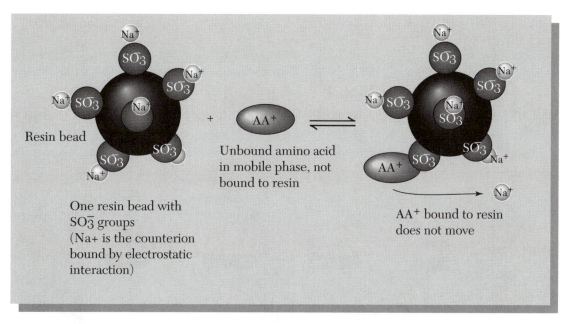

Figure 5.2 Ion exchange chromatography on Dowex

neutral, so it also passes through the column quickly, but not as quickly as the aspartic acid because it has no component with a net negative charge. The histidine has a net positive charge and is attracted to the column, so it moves the slowest. In fact, it would be stuck to the column unless you changed the pH of the buffer to something higher.

A common way to separate amino acids on such a column is to elute them with buffers of differing pH. If you know the pI of the amino acid and pH of the buffer, you can predict whether it will have a net positive, negative, or neutral charge.

> **If pH > pI, the molecule is negative.**
>
> **If pH < pI, the molecule is positive.**
>
> **If pH = pI, the molecule is neutral.**

A more subtle way of eluting amino acids would be to use a pH gradient. Instead of using different buffers of different pHs, a gradient maker can be used to create a buffer of continually increasing pH. In this way, amino acids with small differences in pI can also be separated. Sometimes it is undesirable to change the pH of the buffers. It is also possible to elute amino acids or other charged molecules by adding excesses of the same charge. For instance, by adding a high concentration of sodium ions to the Dowex column described earlier, the amino acids could be eluted because the sodium ion outcompetes them for the sulfonyl sites.

▣➡ Practice Session 5.5

If we have a mixture of Phe, pI 5.5; Thr, PI 6.5; Asp, PI 3; His, PI 7.6, and Lys, PI 9.8, at a pH of 11 and we load it on an anion exchange column and then elute with a pH gradient of lowering pH, in what order will the amino acids elute?

(a)

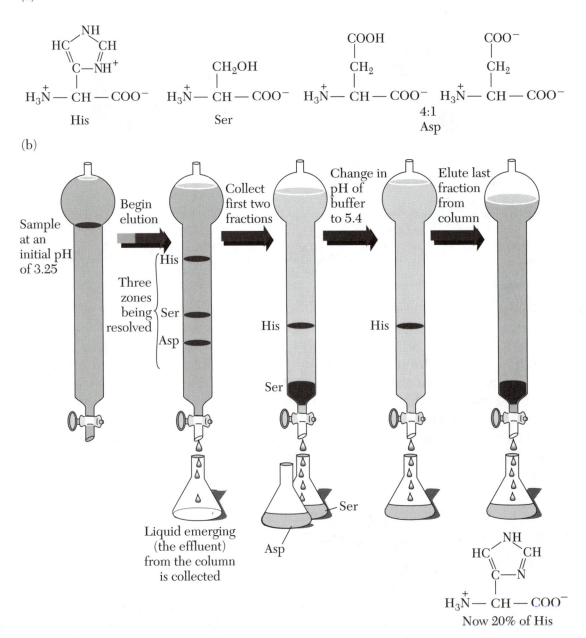

(b)

Figure 5.3 Elution of amino acids with ion exchange chromatography

At pH 11, all will be negatively charged because all of the pI's are less than 11. The amino acids will elute in order as their pIs are reached, so the order will be Lys, His, Thr, Phe, Asp.

5.4   Ion Exchange Resins

Ion exchange resins are made up of two parts. The first is the insoluble, three-dimensional matrix or support. The second is the chemically bonded charged group that is responsible for the ion exchange effect. The gel matrix can be made from a variety of materials. Common

Table 5-2 Typical Ion Exchange Resins

	Functional Group	Matrix	Class
Anion Exchangers			
AG 3	Tertiary amine	Polystyrene	Weak
DEAE-cellulose	Diethylaminoethyl	Cellulose	Weak
DEAE-Sephadex	Diethylaminoethyl	Dextran	Weak
QAE-Sephadex	Diethyl-(2-hydroxyl-propyl)-aminoethyl	Dextran	Strong
Q-Sepharose	Tetramethyl amine	Agarose	Strong
Cation Exchangers			
Dowex	Sulfonic acid	Polystyrene	Strong
CM-cellulose	Carboxymethyl	Cellulose	Weak

ones are polystyrene resins, like Dowex, agarose resins, which carry the common trade name Sepharose; dextran resins, which are usually sold as Sephadex™, and polyacrylamide resins, which are usually called Bio Gel™. All of these can have controlled pore sizes that affect the ability of a molecule to enter the inside of the gel matrix. For small molecules like amino acids or small proteins, polystyrenes like Dowex can be used. For larger molecules, like most enzymes, a larger pore size is needed, so dextrans and agarose-based gels are used.

The most important part of an ion exchanger is the charged group that is chemically bonded to the support. A **cation exchanger** is a negatively charged group, such as a sulfonic acid or carboxylic acid. An **anion exchanger** is a positively charged group, such as a substituted amino group. Ion exchangers are also classified based on the ionizing strength of the charged group. A quaternary amine is always charged, as is a sulfonic acid. These would be a strongly basic and strongly acidic ion exchanger, respectively. Other groups ionize to a lesser extent, such as the weak anion exchanger, DEAE-Sephadex, and the weak cation exchanger, CM-Sephadex. Table 5.2 gives some typical ion exchange resins.

When you are selecting an appropriate ion exchange resin, you must first decide whether you need a cation or an anion exchanger. This will depend on what you are trying to purify and the pH you will need to use in the column. If you select an exchange column, then you also need to decide if you want the functional group to be strong or weak. Strong exchangers are usually reserved for sturdier materials, such as amino acids; the weak exchangers are used more for proteins. Many proteins would be denatured in the environment of a strong exchanger.

5.5 Identification of Compounds Eluted from Columns

When you perform ion exchange chromatography, you put a sample of biological molecules on the column. Some of them will stick to the column while others wash off. Then you elute the bound molecules with a salt gradient or a pH change. Either way, you now have to find the eluted molecules and identify them. Very often spectrophotometry with or without a colorimetric assay is used to find the column fractions that contain the molecules of interest.

Detection of Amino Acids

If you have eluted some amino acids from a column, how do you find them? They are colorless, and only a few can be measured spectrophotometrically. To do this we usually react the amino acids with ninhydrin, which gives a purple color with all of the amino acids except proline. Figure 5.4 shows the reaction.

Identification of Proteins

When proteins are separated by ion exchange chromatography, fractions are obtained by collecting drops off the column into different test tubes. Some of the tubes will have the proteins of interest in them. Usually, a UV monitor is attached to the column outlet so that the absorbance at 280 nm can be measured while the column is running. When the absorbance is high, you know that protein is eluting from the column. This would tell you which tubes had protein in them, but in a mixture of proteins you may not have isolated the desired protein. The UV monitor is used to narrow down the choices. Instead of searching for your protein in 200 samples, you may now have limited your search to 20 or 30. Then the job is to assay the likely fractions for your protein by whatever specific assay is appropriate for that protein. When a flow-through UV monitor is not available, fractions could be quickly assayed at 280 nm with a standard UV spectrophotometer. If this is not available either, then

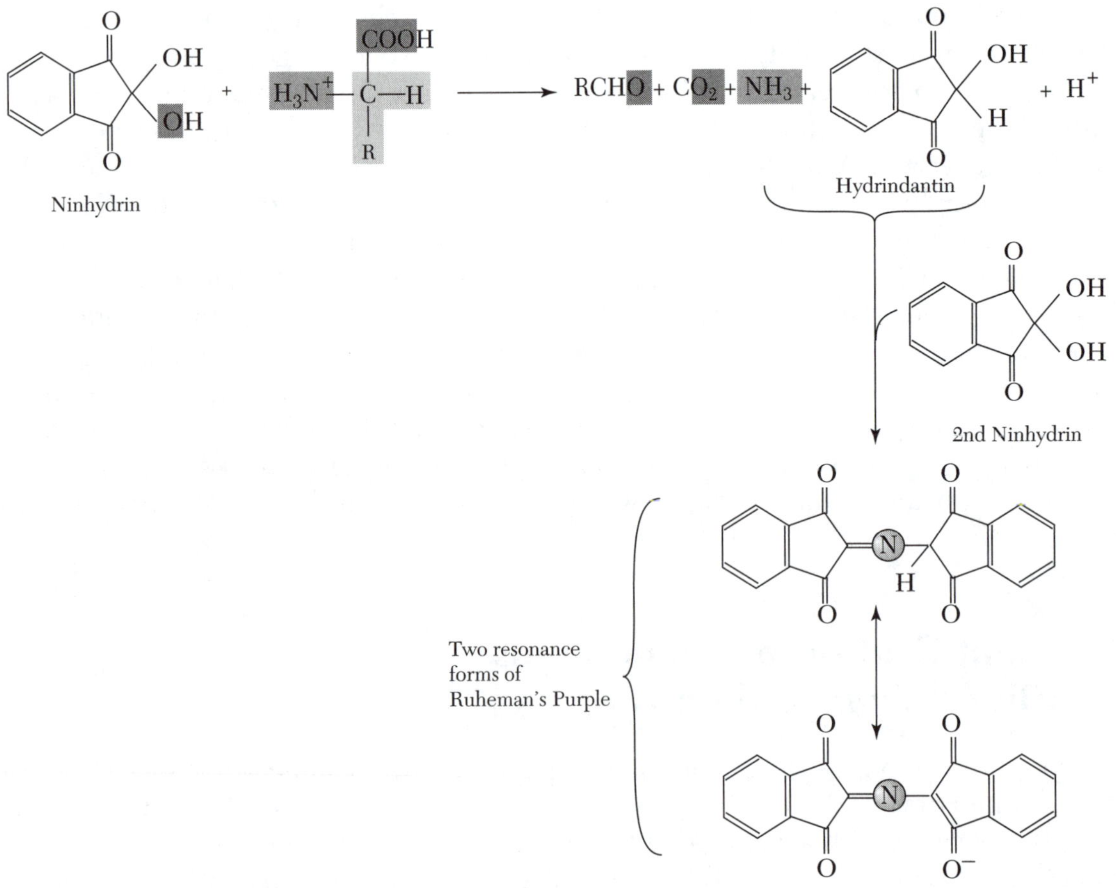

Figure 5.4 The ninhydrin reaction with amino acids

each fraction will have to be assayed for protein using a colorimetric assay, or for the specific enzyme via its enzymatic assay.

5.6   Thin Layer Chromatography

Another technique often used to identify specific amino acids is thin layer chromatography (TLC). TLC is a type of partition chromatography. There is a stationary phase and a mobile phase. Molecules are partitioned between the two, with their affinity for one or the other controlled by the nature of the stationary phase matrix and the mobile phase buffer. Amino acid unknowns are spotted on a thin layer chromatogram along with amino acid standards. The TLC is then developed in a solvent, and the amino acids travel up the chromatogram at characteristic rates dependent mainly on the polarity of the amino acid. Silica gel is very polar, so polar amino acids will tend to stay with the stationary silica phase. Nonpolar amino acids will tend to travel with the mobile phase, which is usually more nonpolar.

The quality of your TLC is based primarily on how well you apply the samples. Very small sample spots must be applied to the thin layer resin because they will spread as the chromatogram develops. The quality of the chromatogram plate and the care with which you put it into the solvent are also very important.

Rf Values

Have you ever wondered why we calculate retention factors, Rf's, for many types of chromatography? Whether it is called an Rf, as in this experiment, or an Rm (relative mobility), as in many types of gel electrophoresis, it is common to divide the distance a sample migrated by the distance of a solvent front of some kind. There are two reasons for doing this.

1.   Non-linear Solvent Front. If the solvent front is not straight, such as in Figure 5.5, you could have the possibility that two spots that ran the same distance were not, in fact, the same compound. Since the solvent front ran differently, they appear to be the same but are not. Calculating an Rf will give different numbers for these two samples.

Figure 5.5   Thin layer chromatography (TLC) with an irregular solvent front

2. Reproducibility Between Experimenters. If we run these amino acids in the same solvent on the same thin layer support, we may or may not get the same distances. If our samples run half as long as yours do, the numbers will not be comparable. However, by calculating Rfs, we should get ratios very similar.

5.7 Why Is This Important?

The two techniques we will learn are very important to most of the life sciences. Ion exchange chromatography is used in every subfield of biochemistry and related sciences. By altering the exchange resin, we can control the types of molecules that bind. We can purify a component by binding the impurities or by binding the component of interest and letting the impurities wash through. Two different columns could be used that would accomplish both. A popular separation technique that allows several components to be resolved in a few minutes is high-performance liquid chromatography (HPLC). Many of the columns used with HPLC are just fancy ion exchange columns. When you take away all of the fancy peripherals, you get back to basic ion exchange chromatography. This type of chromatography is very common these days with DNA purification. The costly and somewhat dangerous technique of cesium chloride centrifugation has largely been replaced by small ion exchange columns. Where plasmid preps used to take hours, they can now be done in minutes with the Qiagen® ion exchange system.

Thin layer chromatography is equally useful. As you will see, good separations can be attained in a short time. Different matrices exist for the separation of all kinds of biomolecules. Many complicated detection systems employ TLC. Drug analysis is just one of them. The preliminary screen of an athlete's urine sample is usually done with a fancy TLC plate.

Experiment 5

Separation and Identification of Amino Acids

In this experiment, you will conduct an analysis of a solution containing several amino acids. The amino acids will be separated by cation exchange chromatography and then identified by TLC. If you plan ahead and you and your lab partner work efficiently together, you can be done on time.

Prelab Questions

1. In part B, step 3, we ask you to test the effectiveness and specificity of ninhydrin by spraying spots of water, bovine serum albumin (a protein), and amino acids with ninhydrin. What do you predict will be the results and why? What assumptions must you make?

2. Summarize in a couple of sentences what we are going to do in this experiment, why we are using two different techniques, and the information we will get from each.

Objectives

Upon successful completion of this experiment, the student will be able to

1. Predict the dominant ionic form of an amino acid at a given pH.
2. Calculate the percentage of the amino acid in the various forms at a given pH.
3. Explain how anion and cation exchange columns work.
4. Predict the order of elution of amino acids off an ion exchange column.
5. Separate a mixture of amino acids into acidic, basic, and neutral amino acids using a Dowex 50 column.
6. Apply samples properly to columns and thin layer chromatograms.
7. Identify amino acids by thin layer chromatography using suitable standards.

Experimental Procedures

Materials

Dowex 50 $\times$ 8 (200–400 mesh) in 0.05 M citrate, pH 3

Citrate buffer, 0.05 M, pH 3 and 6

CAPS buffer, pH 11, 0.05 M

Glass columns

Amino acid mixtures, 1% in pH 3 citrate (possibilities will be on the blackboard)

Ninhydrin solution

Whatman™ 3 MM chromatography paper, 3 × 10 cm

TLC plates, 7 × 10 cm

Propanol/acetic acid/water solvent, 4/1/1

Drawn-out capillary tubes for spotting paper chromatograms

Amino acid standards, 2% w/v in water

Procedures

Part A: Separation and Identification of Amino Acids by TLC

1. Obtain a 7- × 10-cm thin layer chromatogram. Handle it only with gloves, and do not touch the gel surface. The chromatogram is prepared by spotting along one edge.

2. Draw a line on a piece of paper. Set the chromatogram on the paper gel side up, such that the line is 1.5 cm from the bottom. *Do not* write on the chromatogram itself.

3. Make room for seven spots—six amino acid standards and the original unknown. Prepare a key to indicate which amino acids are where on your paper.

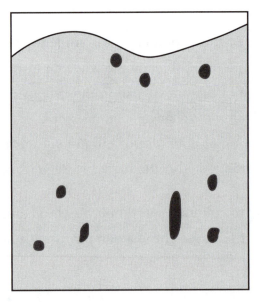

Chromatogram after marking the solvent front and spraying with ninhydrin

4. Using drawn out capillary tubes, spot one application of the amino acids. Make the spots as small as possible (0.2 cm).

5. After all of the spots have dried, carefully place the chromatogram into a beaker with 1 cm of PAW solvent in the bottom. Put the chromatogram in straight with the samples at the bottom. The solvent must be below the line of samples. Cover the beaker with aluminum foil. Proceed to part B.

6. Let the solvent proceed until it is three-fourths of the way to the top. Use a pencil to mark where the solvent front was when you took it out of the solvent.

7. Spray the chromatogram with ninhydrin in the hood (*wear gloves*), and develop it in an oven (110°C) for 10 min or so, or carefully dry it with a hair dryer (not too hot).

Part B: Separation of Amino Acid Mixtures by Ion Exchange

1. Obtain an unknown amino acid mixture from the stockroom.

2. Obtain a Dowex 50 column. These have been prepared in the sodium form and are equilibrated in 0.05 M citrate pH 3.

3. To check the ninhydrin reaction, take one strip and place one spot of water, one spot of your original unknown, and one spot of the protein, BSA. Spray the strip with ninhydrin, heat it in an oven or with a hair dryer and compare the color and intensity of the spots.

4. Remove the plug on the column, allow the buffer on top of the column to drain until the meniscus is just above the resin, and replug the column.

5. From now on, ***do not allow the column to run dry!*** Always use a Pasteur pipet.

6. Prepare 30 or so test tubes for fraction collecting by marking with a line at a 1-mL level.

7. Using a Pasteur pipet, carefully apply 0.5 mL of amino acid unknown to the top of the resin. Do not disturb the resin.

8. Collect drops into tube 1.

9. When the meniscus has just entered the resin, replug the column, and add 1 mL of citrate pH 3 buffer carefully. Wash the sides of the column. Continue collecting.

10. When 1 mL has been collected in tube 1, change to tube 2.

11. When the buffer has entered the resin, carefully add 5–10 mL of buffer, and continue collecting 1 mL fractions.

12. To test for the presence of an amino acid, take 1 drop of the fraction and apply to the 3- × 10-cm strip of paper. Spray the strip with ninhydrin in the hood (*wear gloves*), and dry it with a hair dryer. For each buffer you elute with, save the strip that turns the darkest purple.

13. Continue the collection until the fractions no longer turn purple. Alternatively, if you have collected 20 mL and still see no purple color, then no amino acids are coming off at that pH.

14. Drain and/or remove the buffer from the top and replace carefully with 1 mL citrate pH 6. Repeat steps 8–12 with this buffer until again no more amino acids come off.

15. Repeat with 0.05 M CAPS, pH 11.

Name ______________________________ Section __________

Lab partner(s) ______________________________ Date __________

Analysis of Results

Experiment 5: Separation and Identification of Amino Acids

Data

Unknown # __________

Part A

1. Draw a picture of your TLC results in the box. You may also attach your TLC if you wish.

2. Make sure you label the drawing so that it is clear which amino acids are in each column.

3. Calculate the Rf for each spot on your chromatogram. The Rf is the ratio of the distance the spot moved to the distance the solvent moved.

Amino Acid	Rf
_____	_____
_____	_____
_____	_____
_____	_____
_____	_____
Unknown	_____

Part B

1. Of the three pH buffers that you used, which ones caused the elution of amino acids?

2. Attach your paper strips to this report as justification for your answer to Question 1.

Analysis and Questions

1. Where should the amino acids we used elute from the column, that is, with which buffers?

Amino Acid	Buffer It Should Elute With
_____	_____
_____	_____
_____	_____
_____	_____
_____	_____

2. Based solely on the ion exchange column, which amino acids could you have in your unknown?

3. Identify the amino acids in your unknown. If you cannot decide between possibilities, at least narrow down the possibilities. Explain.

4. Determine the isoelectric point of the following peptide:

   H_3N^+-Asp-Lys-Leu-Tyr-Glu-His-COOH

   Show your work on how you calculated your answer.

5. Draw all the forms of histidine at pH 6 present at a level of at least 10% of the total possible forms.

Experiment 5a

Purification of LDH with Ion Exchange Chromatography

In this experiment, you will use ion exchange chromatography with Q-Sepharose to continue your purification of LDH that you started in Experiment 4a.

Prelab Questions

1. What is the chemical nature of Q-Sepharose that allows it to be used as an ion exchanger?
2. What is the relationship between the pI of LDH and the pH of the buffers we will be using?
3. List three ways of eluting LDH from the column once it is bound.

Objectives

Upon successful completion of this experiment, the student will be able to

1. Properly pour an ion exchange column.
2. Properly set up the column, fraction collectors, and UV monitors.
3. Properly load the LDH sample on the column and elute with appropriate buffers.
4. Explain how anion and cation exchange columns work.
5. Analyze UV monitor results and predict location of LDH.
6. Assay fractions and determine recovery from column.

Experimental Procedures

Preparation of IEX Columns

Materials

Bio Rad Econo® columns (1.5 × 15 cm)

Fast Flow Q-Sepharose equilibrated in 0.03 M bicine, pH 8.5

Plugs, caps, tubing, and connectors for columns

Procedures

All we are doing today is getting the first column ready that you will use after the tissue preparation phase is completed.

1.   Acquire some Q-Sepharose. This should be degassing when you arrive.

2.   Acquire a 15- × 1.5-cm Bio Rad Econo column, ring stand, and clamp.

3.   Pour and pack the column until the packed resin is about 2 cm below the yellow top (i.e. still visible in the clear glass).

4.   Make sure the column is well-stoppered at the bottom and top. Parafilm both the bottom and top to help keep it from leaking.

5.   Place the columns in a designated location.

Ion Exchange Chromatography

Materials

Q-Sepharose columns

Fraction collectors

UV monitors

0.03 M bicine, pH 8.5

CAPS buffer, 0.14 M, pH 10

NAD^+, 6 mM

Lactate, 0.15 M

Bicine, 0.03 M, pH 8.5 (wash buffer)

Bicine, 0.03 M, pH 8.5, with 0.2, 0.4, 0.6, 0.8, or 1 M NaCl added (elution buffer)

0.02 M sodium phosphate, pH 6.0

Procedures

1.   Collect the ion exchange columns that you prepared previously. Collect your dialysis bags, keeping them cool until needed.

2.   If you have a visible precipitate in the dialysis bag, transfer the solution to a centrifuge tube and spin for a couple of minutes using a benchtop clinical centrifuge. Save the supernatant.

3.   Save a small aliquot of the dialyzed LDH for assay. It will be very important to know how many units you are loading onto the column. This assay should be done early in the day. In addition, this will tell you if you lost any activity during the dialysis.

4.   Make sure the column, fraction collector, and UV monitor are all assembled properly, turned on, and warmed up. If any of these are not available, the procedure below will have to be modified.

Ion Exchange Loading

5. Drain the excess buffer off the top of the column by allowing the column to run with the fraction collector tube dripping into a graduated cylinder. Set the fraction collector to collect drops and set the counter to its maximum value so that it will not advance during this procedure. Figure out how many drops it takes to give you 1 mL. This is best done by collecting 5–10 mL into the graduated cylinder so that you get a better average.

6. When the resin is just about to run dry at the top, stop the column. Have the fraction collector ready to collect drops and set the volume to 3 mL.

7. Slowly load the dialyzed LDH into the column, being careful not to disturb the bed. After loading, allow the column to flow. Be very careful that it doesn't run dry. Once the last of the LDH is about to disappear into the resin, stop the column.

Ion Exchange Wash

8. Add 0.03 M bicine, pH 8.5, and resume collecting. *Make sure the column never runs dry.*

9. Collect fractions until the UV monitor spikes and then goes back to baseline. Mark the tubes periodically so that you know what tubes are being collected at various positions on the chart recorder. If no UV monitor is available, you may have to assay fractions with another type of UV spectrophotometer.

10. Assay the fractions that have significant protein as indicated by the UV monitor.

11. If any of these fractions have significant LDH activity, calculate how many units have been collected.

Ion Exchange Elution

12. If it appears that most of your LDH bound to the column, then you will need to elute it. This can be done with a batchwise application or a salt gradient.

13. When the last of the wash buffer is disappearing into the resin, switch over to elution buffer. You will start with 15 mL of bicine + 0.2 M NaCl. If you have a gradient maker available, you will set it up going from plain bicine to bicine plus 1 M NaCl, 40 mL each per chamber. Step 15 would then be irrelevant.

14. Collect fractions as before, watching the UV monitor carefully for evidence of peaks eluting. Hopefully, your LDH will elute with a peak, as that makes it easier to find.

15. When you are out of the 0.2 M NaCl elution buffer, switch to the 0.4 M NaCl buffer, and so on.

16. Find the LDH and pool the fractions with the most significant activity. Note, if you pool fewer fractions, you will have less material to work with, but it may be more pure. There are still two columns to go to attain reasonable purity, so it is best to optimize for yield rather than purity at this step.

17. Assay the pooled fractions and save some for protein determination as always.

18. Place the pooled fractions into a dialysis bag and put them into a bucket of 0.02 M sodium phosphate, pH 6.0, that is at 4°C.

Analysis of Results

Experiment 5a: Purification of LDH with Ion Exchange Chromatography

Data

1. Fill in the following table concerning the dialyzed sample you loaded on the column.

Data from Dialyzed Sample

μL assayed	
Dilution used (if any)	
Δ Absorbance/Δ min	
Units/mL	
Total mL loaded on column	
Total units loaded on column	

2. Fill in the following table for the wash buffer fractions.

Data from Wash Fractions

Fraction #	Volume Assayed	$\Delta A/\Delta$ min	Units/mL	Total Units

3. Fill in the following table for the elution fractions.

Data for Elution Fractions

Fraction #	Volume Assayed	$\Delta A/\Delta$ min	Units/mL	Total Units

Analysis of Results

1. Fill in the following table concerning your IEX results.

Summary of Ion Exchange Results

Fractions pooled	
Volume assayed	
Δ Absorbance/Δ min	
Units/mL	
Total units	
% Recovery off of column	
% Recovery from beginning of experiment	

2. What percentage of the loaded activity eluted in the wash buffer?

3. What percentage eluted in the elution buffers?

4. With which salt concentration did the most activity elute? If you used a gradient, estimate the salt concentration from the volume, assuming a linear change from zero to 1 M.

5. Was there a significant difference between the relative activity of the 65% pellet from Experiment 4a and the dialyzed pellet you used today? If so, how much activity did you lose? How could this be improved?

Additional Problem Set

1. Write equations to show the ionic dissociations of the following amino acids: glutamic acid, isoleucine, histidine, lysine.

2. Predict the most prevalent form of the following amino acids at pH 7: glutamic acid, isoleucine, histidine, arginine, aspartic acid, cysteine.

3. Based on the information in Table 5.1, are there any amino acids that could make good buffers at pH 8.0? Which ones? Which one would be the best buffer?

4. Given the following peptide:

    Val-Met-Ser-Gly-Glu-Ser-Asp-His-Lys-Cys-Tyr-Leu

    a. What is the pI?
    b. What would be the net charge at pH 7.0?
    c. What would be the net charge at pH 4.0?

5. Consider the following peptides:

    Phe-Glu-Ser-Met   and   Val-Trp-Cys-Leu

    Do these peptides have different net charges at pH 1.0? At pH 7.0? Indicate the charges at both pH values.

6. A solution containing aspartic acid, glycine, threonine, leucine, and lysine is applied to a Dowex 50 cation exchange column at pH 3.0. If the amino acids are eluted with an increasing pH gradient, in what order will they elute?

7. To determine the isoelectric point of an amino acid or a peptide, why is it important to know both the structure and the pK_a of the functional group instead of just the pK_a?

8. Why must the pI be an average of two pK_as?

9. What is the most useful definition of acid and base when discussing the side chains of amino acids and their pIs?

10. What are the three amino acids that would be considered basic according to the definition in problem 9?

Webconnections

For a list of Web sites related to the material covered in this chapter, go to **Webconnections** at the *Experiments in Biochemistry* site on the Saunders College Publishing Web page. You can access this page at http://www.saunderscollege.com. **Webconnections** are under *Experiments in Biochemistry* in the Biochemistry portion of the Chemistry page.

References and Further Reading

H. Ahern, (1996), *The Scientist,* **10(5).** Chromatography.
R. F. Boyer, *Modern Experimental Biochemistry,* Addison-Wesley, 1993.
A. Braithwaite and F .J. Smith, *Chromatographic Methods,* Blackie Academic and Professional, 1996.

M. Campbell, *Biochemistry,* Saunders College, 1998.

T. G. Cooper, *The Tools of Biochemistry,* Wiley Interscience, 1977.

P. B. Hamilton, (1963), *Anal. Chem.* **35(13).** *Ion Exchange Chromatography of Amino Acids: A Single Column, High Resolving, Fully Automatic Procedure.*

K. Robards, P. R. Haddad, and P. E. Jackson, *Principles and Practice of Modern Chromatographic Methods,* Academic Press, 1994.

J. F. Robyt and B. J. White, *Biochemical Techniques,* Waveland Press, 1990.

I. H. Segel, *Biochemical Calculations,* Wiley Interscience Publishers, 1976.

T. Shihamoto, *Chromatographic Analysis of Environmental and Food Toxicants,* Marcel Dekker, 1998.

G. T. Tsao, *Chromatography,* Elsevier Science, 1991.

Chapter 6

Affinity Chromatography

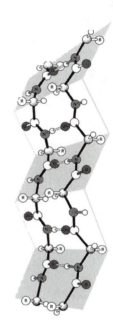

Topics

Introduction

This chapter discusses another type of adsorption chromatography, affinity chromatography, which is based on the interactions between a protein of interest and a ligand bound to the gel matrix. Because the binding of the proteins is specific to the structure of a bound ligand, excellent purifications can be achieved with this technique.

6.1 Affinity Chromatography

Affinity chromatography is a type of adsorption chromatography, similar to ion exchange. With ion exchange, however, the basis of the separation of molecules was the charge nature of the protein compared to the gel. With a cation exchanger, for example, any protein with a net positive charge would be expected to bind. With affinity chromatography, the nature of the binding is more specific. We design or buy a column resin that has a ligand on it that our molecule recognizes. For the separation of LDH, for example, an affinity resin might resemble one of the substrates, either lactate, pyruvate, NAD^+, or NADH. Figure 6.1 demonstrates the basic principles.

A mixture of proteins is passed over the affinity column. The affinity ligand has a binding affinity for a protein of interest, which

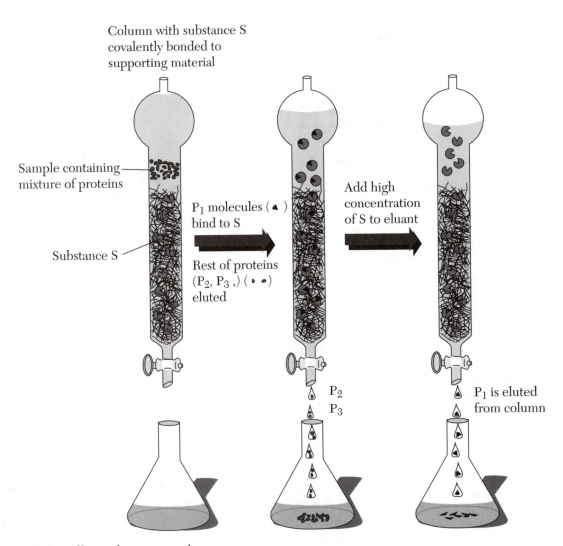

Column with substance S
covalently bonded to
supporting material

Sample containing
mixture of proteins

Substance S

P_1 molecules (▲)
bind to S

Rest of proteins
$(P_2, P_3,)$ (● ●)
eluted

Add high
concentration
of S to eluant

P_2
P_3

P_1 is eluted
from column

Figure 6.1   Affinity chromatography

binds to the ligand. The other proteins elute from the column in the wash. Then the bound protein of interest is eluted.

   Affinity chromatography can give extensive purifications in a single step due to the specific nature of the binding. In theory, you could achieve complete separation with this step if the binding between the ligand and protein were specific enough. This rarely happens in practice, of course.

6.2   Gel Supports

The gel supports used in affinity chromatography are the same ones we discussed in the last chapter. Most affinity gels are created by covalently linking the ligand to a matrix of agarose, dextran, or polyacrylamide. In the experiment that follows, we use Cibacron Blue Sepharose. The Sepharose is the support. The Cibacron Blue is the ligand.

6.3 Affinity Ligands

The ligands used for affinity chromatography can be broken down into two broad categories—individual and group-specific.

Individual Ligands

An individual ligand is one that is very specific for a protein of interest. For example, Figure 6.2 shows the structure of NAD^+. If we made or bought an affinity resin that was composed of NAD^+ attached to a Sephadex matrix, we would have a very specific affinity resin for LDH because NAD^+ is one of the substrates for the enzyme. The binding between the ligand and the enzyme would be expected to be very good, although it would not be absolutely specific because other enzymes use NAD^+ as substrates.

Group-Specific Ligands

A group-specific ligand is a little less specific in its interaction than an individual ligand. A good example is AMP-agarose. AMP has the structure shown in Figure 6.3. If you look at AMP carefully, you can see that it is essentially half of NAD^+. AMP-agarose would be

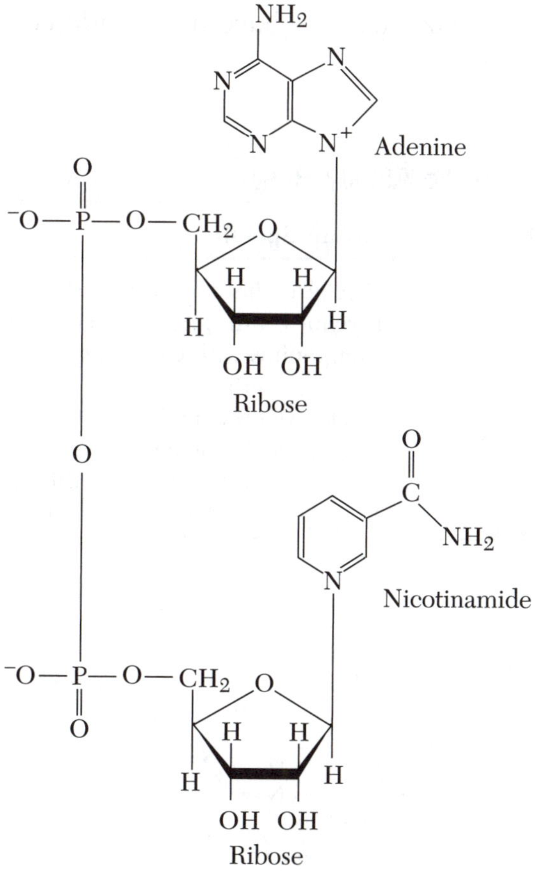

Figure 6.2 Structure of NAD^+

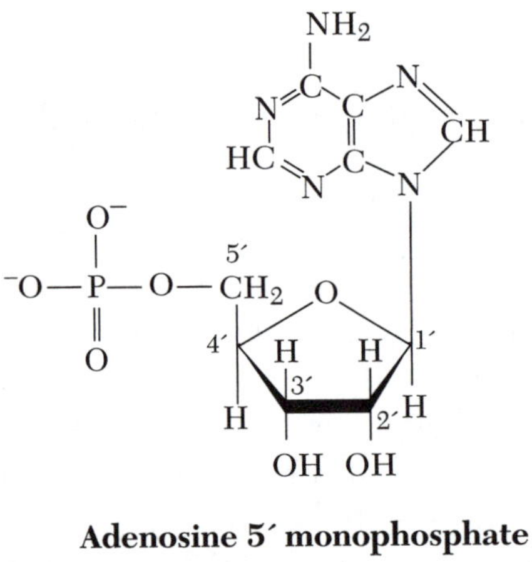

Adenosine 5′ monophosphate

Figure 6.3 Structure of AMP

expected to bind enzymes that use AMP directly or enzymes that bind NAD^+ cofactors, since the AMP would fit nicely into the active site of any molecule that did use NAD^+. Table 6.1 gives some examples of group-specific resins.

Some of the binding properties are obvious, such as with the AMP-agarose, but there are group-specific resins that are not as obvious from their name or structure. Figure 6.4 shows the structure of Cibacron Blue. It is a dye that is linked to agarose-type resins and makes an excellent affinity ligand for enzymes that have nucleotide cofactors, which include the dehydrogenase family of enzymes.

Table 6.1 Group-Specific Affinity Resins

Group-Specific Adsorbent	Group Specificity
Concanavalin A–agarose	Glycoproteins and glycolipids
Cibacron Blue–agarose	Enzymes with nucleotide cofactors
Boronic acid–agarose	Compounds with Cis-diol groups
Protein A–agarose	IgG-type antibodies
Poly(U)-agarose	Nucleic acids containing poly(A) sequences
Poly(A)-agarose	Nucleic acids containing poly(U) sequences
Iminodiacetate-agarose	Proteins with heavy metal affinity
5′-AMP–agarose	Enzymes with NAD^+ cofactors, ATP-dependent kinases

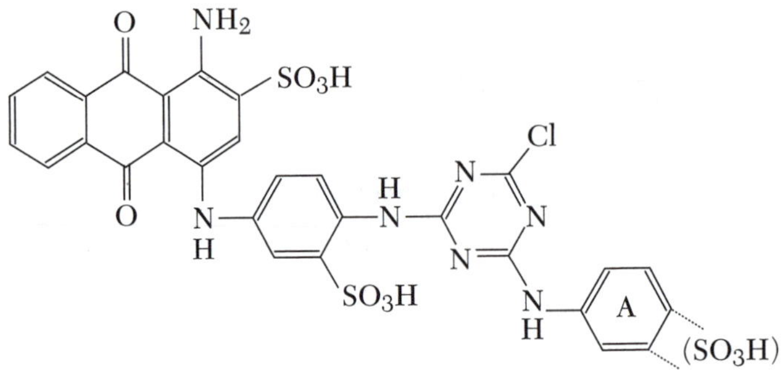

Figure 6.4 Structure of Cibacron Blue 3G

6.4   Elution of Bound Molecules

Once a protein of interest has bound to the affinity resin, you must find a way to elute it. There are two basic plans for doing this.

High-Salt Elution

Just as we saw with ion exchange chromatography, adding a solution of high salt concentration will often disrupt the binding of the ligand and the protein. Although the binding was not based on net charge, almost all binding of enzyme to substrates is based upon electrostatic interactions. A solution with very high ionic strength can disrupt those interactions and cause the bound molecule to be released. This can be done in batch style or with a gradient just as in ion exchange chromatography.

Affinity Elution

This is the type of elution shown in Figure 6.1. Instead of eluting the bound protein with high salt, the bound protein is eluted with its own substrate. If you had LDH bound to an AMP-agarose column, running NADH over the column would likely remove the LDH, because the LDH would recognize its own substrate and bind to it rather than the column. The advantage to affinity elution is that it gives another level of specificity to the separation. The disadvantage is that the substrates used to affinity elute are often very expensive.

> **Tip 6.1**
> ⊃ As with any chromatography step, the challenge is often to find the enzyme once you elute. Be sure to measure the number of units you load on the column and compare that to the number of units that elute.

6.5   Why Is This Important?

Of all the column purification techniques, affinity chromatography has the greatest potential for increases in purity with a single step. An enzyme for which you could create a very specific ligand would be easy to purify. This approach is currently being used in molecular biology, where genes are being modified so that the protein they express contain a specific tag, such as an N-terminal series of histidines. An affinity column based on chelated nickel ions will bind the histidines very tightly and bind nothing else. Nickel column manufacturers claim that histidine-tagged proteins can be separated from crude homogenates to complete purity in only one step.

Experiment 6

Affinity Chromatography of LDH

In this experiment, you will use affinity chromatography with Cibacron Blue Sepharose to purify lactate dehydrogenase. This would work well after the ion exchange chromatography from Experiment 5a, or it could be a separate experiment.

Objectives

Upon successful completion of this experiment, the student will be able to

1. Prepare a Cibacron Blue Sepharose column.

2. Assay a sample for LDH, calculate relative activity, and calculate the number of units loaded on the column.

3. Wash unbound proteins from the column and monitor with a UV flow-through spectrophotometer.

4. Elute LDH using NaCl and isolate the active fractions.

5. Pool and dialyze elution fractions with significant activity and calculate yield.

6. Explain how Cibacron Blue works to purify LDH.

Experimental Procedures

Preparation of Cibacron Blue Sepharose

Materials

    1.5- × 15-cm Econo columns and accessories

    Cibacron Blue Sepharose in 0.02 M sodium phosphate, pH 6.0

Procedures

1. Acquire the columns and accessories you will need.

2. Acquire the column resin you will need.

3. Degas the Cibacron Blue Sepharose with a sidearm flask and vacuum line until no more bubbles are coming off.

4. Pour the Cibacron Blue Sepharose to a height of about 10 cm.

5. Stopper all columns well and wrap both top and bottom with Parafilm. Store in the usual location until needed.

Affinity Chromatography of LDH

Materials

1.5- × 15-cm Econo columns

Cibacron Blue Sepharose in 0.02 M sodium phosphate, pH 6.0

Fraction collectors

UV monitors

CAPS buffer, 0.14 M, pH 10

NAD^+, 6 mM

Lactate, 0.15 M

0.02 M sodium phosphate, pH 6.0 (wash buffer)

Sodium phosphate, pH 6.0 with 0.2 M, 0.4 M, 0.6 M, 0.8 M, and 1 M NaCl (elution buffers), or just the 1 M NaCl if you will use a gradient

Millipore spin columns

Procedures

1. Acquire your affinity column and dialyzed LDH sample from Experiment 5a. If this experiment is being done apart from the LDH purification series, you will be using another LDH sample for this and step 2 will be unnecessary.

2. If a precipitate has occurred, spin it down as before.

3. Set up the fraction collectors, UV monitors, and columns used in Experiment 5a. If the tubing is the same size, you should not need to recalibrate the drop count vs. volume. If you are not using a UV monitor or fraction collector, these procedures will have to be modified.

4. Assay your LDH fraction to see if any activity has been lost during dialysis and to see how much you are loading on the Cibacron Blue Sepharose column.

5. Load the sample and collect 5-mL fractions. When the sample is loaded, switch to the 0.02 M sodium phosphate wash buffer. Continue washing until the UV monitor is back to baseline or you have otherwise determined that all of the protein that is not sticking to the column has washed off.

6. Assay the wash fractions to be sure your LDH bound.

7. Elute with 10-mL aliquots of the sodium phosphate plus NaCl as before, collecting 2-mL fractions, or use a salt gradient.

8. Assay the fractions, isolating the peak fractions with significant activity. Again, the UV monitor may be able to help you here.

9. Pool the peak fractions that contain significant activity.

10. Concentrate your pooled fractions down to 0.5 mL using spin columns. With Millipore Biomax® columns, you can put up to 15 mL in the tubes. Centrifuge at $2000 \times g$ until the solution is concentrated. This will take 15–60 min depending on original concentration and temperature.

11. Assay your concentrated fraction. Note that this should take a sizable dilution if everything went according to plan.

12. Store your samples at 4°C.

Analysis of Results

Experiment 6: Affinity Chromatography of LDH

Data

1. Fill in the following table concerning the dialyzed sample you loaded on the column.

Summary of Dialyzed Sample Loaded onto Cibacron Blue Column

Quantity assayed (μL)	
Dilution used (if any)	
Δ Absorbance/Δ min	
Units/mL	
Total mL loaded on column	
Total units loaded on column	

2. Fill in the following table for the wash buffer fractions:

Data from Wash Fractions

Fraction #	Volume Assayed	$\Delta A/\Delta$ min	Units/mL	Total Units

3. Fill in the following table for the elution fractions:

Data for Elution Fractions

Fraction #	Volume Assayed	Δ A/Δ min	Units/mL	Total Units

Analysis of Results

1. Fill in the following table concerning your affinity results.

Summary of Affinity Chromatography Results

Fractions pooled	
Volume assayed	
Δ Absorbance/Δ min	
Units/mL	
Total units	
% Recovery off of column	
% Recovery from beginning of experiment	

2. What percentage of the loaded activity eluted in the wash buffer?

3. What percentage eluted in the elution buffers?

4. With which salt concentration did the most activity elute? If you used a gradient, estimate the salt concentration from the volume assuming a linear change from zero to 1 M.

5. Was there a significant difference between the relative activity of the pooled IEX fractions from Experiment 5a and the dialyzed pellet you used today? If so, how much activity did you lose? How could this be improved?

Additional Problem Set

1. List three enzymes that you would expect to bind to Cibacron Blue Sepharose.

2. Compare the advantages and disadvantages of using AMP-Sepharose vs. Cibacron Blue Sepharose for the purification of LDH.

3. Bovine LDH has five different isozymes that differ by net charge. How would the charge differences affect its elution from an affinity column? From an ion exchange column?

4. What are three ways to elute a bound molecule from an affinity column? What are the advantages and disadvantages of each?

5. NADH works well for affinity elution of bound LDH. What would be two disadvantages for using it for this purpose?

6. If NAD^+ and NADH could both be used for affinity elution of LDH, which one would you choose and why?

7. You are purifying LDH from a resuspended and dialyzed 65% ammonium sulfate pellet using Cibacron Blue Sepharose chromatography. You load 3 mL of the dialyzed pellet on the column and wash with 50 mL of 0.02 M sodium phosphate, pH 6.0, collecting 5-mL fractions. You then switch to an elution buffer containing 1 mM NADH and collect 3-mL fractions. You assay for LDH via the standard assay used in Experiment 6 and see these results:
   a. Calculate the number of units and units/mL for each fraction.
   b. What percentage of the LDH loaded on the column was recovered?

Data for Question 7 LDH Purification

Fraction Assayed	Volume of Sample Assayed (μL)	Dilution Used	Δ A/Δ min
Dialyzed 65%	50	100/1	0.508
Wash fraction 1	100	—	0.03
Wash fraction 2	100	—	0.08
Wash fraction 3	100	—	0.10
Wash fraction 4	100	—	0.07
Wash fraction 5	100	—	0.02
Wash fraction 6	100	—	0
Elution fraction 1	100	—	0
Elution fraction 2	100	—	0.15
Elution fraction 3	50	—	0.40
Elution fraction 4	10	—	0.50
Elution fraction 5	20	10/1	0.05
Elution fraction 6	20	10/1	0.16
Elution fraction 7	10	10/1	0.12
Elution fraction 8	20	10/1	0.10
Elution fraction 9	10	—	0.24
Elution fraction 10	50	—	0.15
Elution fraction 11	100	—	0.05
Elution fraction 12	100	—	0

c. Which fractions would you pool together to continue on to another purification step? Justify your answer.

d. What could you do to increase the enzyme yield from this column?

Webconnections

For a list of Web sites related to the material covered in this chapter, go to **Webconnections** at the *Experiments in Biochemistry* site on the Saunders College Publishing Web page. You can access this page at http://www.saunderscollege.com. **Webconnections** are under *Experiments in Biochemistry* in the Biochemistry portion of the Chemistry page.

References and Further Reading

G. E. Bannikova, V. P. Varlamov, M. L. Miroshnichenko, and E. A. Bonch-Osmolovskaya, (1998), *Biochem. Mol. Biol. Int.* **44(2),** Isolation of Thermostable phosphatase from the Hyperthermophilic Archaeon *Thermococcus Pacificus* by Immobilized Metal Affinity Chromatography.

P. C. Bevilacqua, C. X. George, C. E. Samuel, and T. R. Cech, Binding of the Protein Kinase PKR to RNAs with Secondary Structure Defects, *Biochemistry,* **37(18),** 1998.

S. C. Blanchard, D. Fourmy, R. G. Eason, and J. D. Puglisi, 1998. *Biochemistry* **37(21),** rRNA Chemical Groups Required for Aminoglycoside Binding.

R. F. Boyer, *Modern Experimental Biochemistry,* Addison-Wesley, 1993.

M. Campbell, *Biochemistry,* Saunders College, 1998.

F. Canduri, R. J. Ward, W. F. de Azevedo Junior, R. A. Gomes, and R. K. Arni, (1998), *Biochem. Mol. Biol. Int.* **45(4).** Purification and Partial Characterization of Cathepsin D from Porcine Liver Using Affinity Chromatography.

K. R. Dharmawardana and P. E. Bock, (1998), *Biochemistry,* **37(38).** Demonstration of Exosite I-dependent Interactions of Thrombin with Human Factor V and Factor Va Involving the Factor Va Heavy Chain; Analysis by Affinity Chromatography Employing a Novel Method for Active-Site-Selective Immobilization of Serine Proteases.

R. L. Dryer and G. F. Lata, *Experimental Biochemistry,* Oxford University Press, 1989.

C. Eisen, C. Meyer, R. Dressendorfer, C. Strasburger, H. Decker, and M. Wehling, (1996), *Eur. J. Biochem.* **237(2).** Biotin-Labelled and Photoactivatable Aldosterone and Progesterone Derivatives as Ligands for Affinity Chromatography, Fluorescence Immunoassays and Photoaffinity Labelling.

D. M. Glerum and A. Tzagoloff, (1998), *Anal. Biochem.* **260(1).** Affinity Purification of Yeast Cytochrome Oxidase with Biotinylated Subunits 4, 5, or 6.

R. C. Jack, *Basic Biochemical Laboratory Procedures and Computing,* Oxford University Press, 1995.

T. Kline, *Handbook of Affinity Chromatography,* Marcel Dekker, 1993.

N. E. Labrou and Y. D. Clonis, (1997), *Bioprocess Eng.,* **16(3).** Simultaneous Purification of L-Malate Dehydrogenase and L-Lactate Dehydrogenase from Bovine Heart by Biomimetic-Dye Affinity Chromatography.

T. Maier, N. Drapal, M. Thanbichler, and A. Bock, (1998), *Anal. Biochem.* **259(1).** Strep-tag II Affinity Purification.

U. H. Mortensen, H. R. Stennick, and K. Breddam, (1998), *Anal. Biochem.* **258(2).** Reversed-Flow Affinity Elution Applied to the Purification of Carboxypeptidase Y.

K. M. Muller, K. M. Arndt, K. Bauer, and A. Pluckthun, (1998), *Anal. Biochem.* **259(1).** Tandem Immobilized Metal-Ion Affinity Chromatography/Immunoaffinity Purification of His-Tagged Proteins.

Y. Nagata, K, Maeda, and R. K. Scopes, (1992), *Bioseparation* **2.** NADP Linked Alcohol Dehydrogenases from Extreme Thermophiles: Simple Affinity Purification Schemes and Comparitive Properties of the Enzymes from Different Strains.

J. F. Robyt and B. J. White, *Biochemical Techniques,* Waveland Press, 1990.

R. K. Scopes, (1977), *J. Biochem.,* **161.** Affinity Elution Chromatographic Procedures.

S. Wittlin, J. Rosel, and D. R. Stover, (1998), *Eur. J. Biochem.,* **252(3).** One-Step Purification of Cathepsin D by Affinity Chromatography Using Immobilized Propeptide Sequences.

Chapter 7

Gel Filtration Chromatography

Topics

Introduction

This chapter addresses the third and final type of column chromatography that we will use to purify lactate dehydrogenase. Gel filtration, also called molecular sieving and size exclusion chromatography, is a preparative and analytical technique that allows us to purify macromolecules away from others of different sizes. Information about the molecular weight of a protein can also be determined.

7.1   Gel Filtration

Proteins and other macromolecules can be separated based on molecular weight by using a cross-linked porous gel. Gels have various degrees of cross-linking that allow certain sizes of molecules to pass through while others cannot. Thus some gels are good at separating large molecules and others are better for smaller ones. The degree of retardation of a particle is related to the molecular weight and shape. Some of the molecules will be excluded from the gel; others will be included. Those that are excluded will pass through the column *faster* because the distance they have to travel

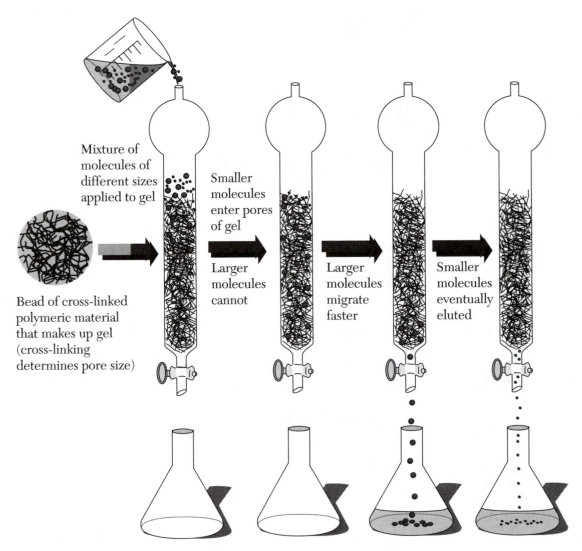

Mixture of
molecules of
different sizes
applied to gel

Smaller
molecules
enter pores
of gel

Larger
molecules
cannot

Larger
molecules
migrate
faster

Smaller
molecules
eventually
eluted

Bead of cross-linked
polymeric material
that makes up gel
(cross-linking
determines pore size)

Figure 7.1   Molecular sieving or gel filtration chromatography

is reduced. The smaller molecules will take a more convoluted path through the column and elute later. Figure 7.1 demonstrates this process.

This technique is useful for many purposes. It is nondestructive, so you do not lose any of your sample. It can be used to determine the molecular weight of a compound in its native conformation. It can be used to purify a compound from other compounds of differing sizes.

7.2   Types of Supports

There are three common types of supports for gel filtration: dextran, polyacrylamide, and agarose. Dextran is a polysaccharide-based resin with glucose residues linked $\alpha\, 1 \rightarrow 6$ with branches of $\alpha\, 1 \rightarrow 3$. These are normally sold under the trade name of Sephadex

Table 7.1 Properties of Sephadex Gels

Dextran Type	Fractionation Range (daltons)	Water Regain (mL/g dry gel)	Bed Volume (mL/g dry gel)
G-10	0–700	1.0	2–3
G-15	0–1500	1.5	2.5–3.5
G-25	1000–5000	2.5	4–6
G-50	1500–30,000	5.0	9–11
G-75	3000–80,000	7.5	12–15
G-100	4000–150,000	10	15–20
G-150	5000–300,000	15	20–30
G-200	5000–600,000	20	30–40

by Pharmacia Chemicals. If you have a bottle of Sephadex, it will say something like G-100. The G refers to the amount of water gained by the dehydrated gel when it is allowed to swell. There are many different sizes of Sephadex, all of which have different levels of cross-linking (i.e. pore size). These will therefore separate molecules in a different range of molecular weights. Table 7.1 gives some properties of the Sephadex gels.

Sephadex gels have different **exclusion limits,** which are the molecular weight limit for entrance into the gel beads. If a gel has an exclusion limit of 80,000, then all proteins of molecular weight 80,000 or greater will not have access to the beads. These limits can be seen as the higher number on the fractionation range in Table 7.1.

Polyacrylamide gels are made of beads composed of the same material used to make acrylamide gels for electrophoresis. These are usually produced under the trade name Bio-Gel by Bio Rad Laboratories.

Agarose beads are made of the same material used for agarose gel electrophoresis. These are usually sold under the trade name Sepharose, also by Pharmacia.

7.3 Determining the Molecular Weight

When native, globular proteins are run on the appropriate size of gel filtration resin, they separate according to their molecular weights. If we plot the log of the molecular weight vs. the elution volume from the column, we should see a straight line. The **elution volume,** V_e, is the volume of liquid that is collected from the moment the sample is applied to the column until the sample is collected from the column. Because each column is different in length, width, particle size, level of hydration, and so on, it would be difficult to compare elution volumes for the same molecule on different columns, even if these columns theoretically had the same material. There are several ways of overcoming this difficulty, all of which are similar to taking an Rf, as we saw with thin layer chromatography. Figure 7.2 shows a typical calibration curve for a gel, using the ratio of the elution volume to the void volume of the column (see Section 7.4). Note that each column should be individually calibrated for the greatest accuracy.

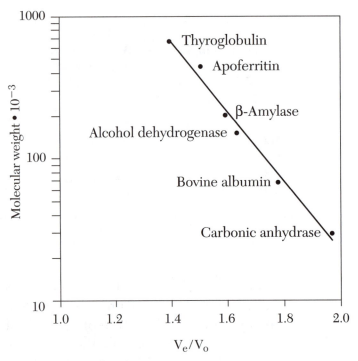

Figure 7.2 Calibration curve for Sephadex gel

7.4 Distribution Coefficients

An equation can be set up for molecules moving through a column:

$$K_d = \frac{V_e - V_0}{V_i}$$

where V_e = **Elution volume for the solute of interest, in this case a protein whose molecular weight we want to determine.**

V_0 = **Void volume, which is the elution volume of a compound that is completely excluded from the gel. This is usually determined by measuring the volume it takes a sample of blue dextran to elute.**

V_i = **Inner volume, the volume of liquid inside the gel bead.**

It is difficult to measure V_i accurately, so the value for the K_d (the real distribution coefficient) is also difficult to measure accurately. A simplifying equation using the total column volume, V_t, is usually used instead.

> V_t = **total column volume, which equals $V_i + V_0 + V_g$, where V_g is the volume of the gel bead itself, not including any solvent. If we assume that the volume of the gel is insignificant compared to the column volume, we can say that $V_t = V_i + V_0$, or, by rearranging, $V_i = V_t - V_0$.**

V_t can be calculated by taking the actual volume of the column as $\pi r^2 h$ or by measuring the amount of water necessary to fill the column to the point where the gel was packed. Assuming that the volume of the gel is negligible will also allow us to calculate the V_t another way. If V_t is considered to be $V_0 + V_i$, then it is also the volume available to a molecule that has access to all of the pores in the gel. A small molecule, such as DNP-aspartate, is used to measure this.

A working coefficient, called K_{avg}, is often used:

$$K_{avg} = \frac{V_e - V_0}{V_t - V_0}$$

This distribution coefficient means that a solute is distributed between the two parts of the column: the space within the porous beads and the space between the porous beads. Graphing log MW vs. K_d or K_{avg} gives a straight line for those proteins falling within the linear separation limits of the gel. Because the difference between K_d and K_{avg} is small and some sources define K_d the same way we defined K_{avg}, we will use K_d throughout to mean K_{avg} as defined earlier.

The K_ds for many proteins have been established for the common types of gel filtration media. To find the molecular weight of an unknown protein, its K_d can be determined, and it can be compared to known K_ds of proteins on the same gel matrix by using a graph. Although this technique is not as accurate as establishing your own calibration curve for your column, it will allow you to estimate the molecular weight of an unknown without running multiple standards.

Because the shape is not taken into consideration, the molecular weight is an estimate. It is quite possible that two proteins that weigh exactly 60,000 would elute differently on the gel if their shapes were quite different. For example, a cigar-shaped protein would have an effective radius much larger than a spherical one of the same weight.

To calculate the molecular weight of an unknown protein by this method, you calculate the V_0, V_e, and V_t of the column. This can be done by running a large molecule, such as blue dextran, to measure the V_0. You would apply the blue dextran to the column and see what volume it takes for it to come off the column. You would then do the same thing with the protein whose K_d you are trying to measure. This would be the V_e. To calculate V_t, you could measure the volume of water it takes to fill the column up to the height of the packed gel. In Experiment 7, we will calculate V_t by using a tiny molecule, a DNP-linked amino acid, to measure the total volume, V_t. When you have run all three of the compounds through the column, you can make a graph such as the one shown in Figure 7.3. The volumes of the three peaks would indicate the void, elution, and total volumes, respectively. The K_d would then be calculated according to the equation given previously.

◁▭▷ Practice Session 7.1

Using the information in the graph in Figure 7.3, what would be the K_d for the protein of interest?

The void volume is the volume at which blue dextran elutes, which is 20 mL. The elution volume is that of the protein, 60 mL, and the total volume is that of the DNP–amino acid, 120 mL. Note that we took the fraction off the column that had the highest absorbance for those compounds to represent the peaks. Thus, the K_d would be calculated as follows:

$$K_d = \frac{(60 - 20)}{(120 - 20)} = 0.4$$

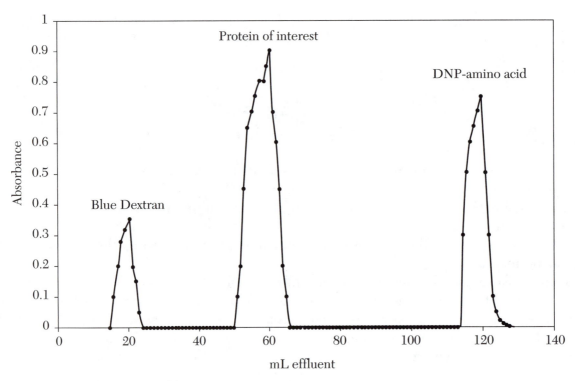

Figure 7.3 Absorbance vs. volume

Practice Session 7.2

What is the MW of the protein of interest?

To answer this question, you must be able to graph log MW of known proteins vs. their K_d values. Assuming you were provided with a table of K_d values and molecular weights, you would be able to make a graph such as the one shown in Figure 7.4. This is a semi-log graph of MW vs. K_d. You just need to interpolate the K_d of your protein of interest off the graph. It just so happens that there is also a known protein with a K_d of 0.4. The MW of a protein with such a K_d is 20,000.

7.5   Why Is This Important?

Gel filtration is a very common, quick, and easy technique that is used in almost every subfield of biochemistry. It is used for purifying proteins, nucleic acids, carbohydrates . . . you name it. It is also used to remove salt or other small molecules as a prelude to continuing purification. For example, if you use ammonium sulfate to salt out a protein, the sample you get then contains a lot of salt. It may be difficult or impossible to put the sample through the next step until the salt is removed. In the experiment that follows, you will see how a large molecule (blue dextran) and a small molecule (DNP-aspartate) separate very quickly on this column. The same would be true for a protein and excess salt.

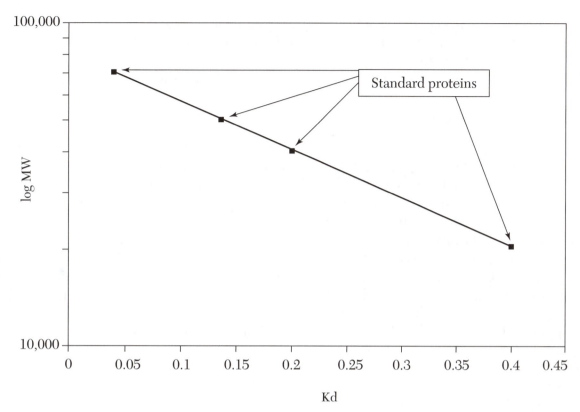

Figure 7.4 log MW vs. K_d

7.6 Expanding the Topic

Calculating Elution Volumes

Many students find it difficult to calculate elution volumes for a couple of reasons. First, we often start a chromatography experiment by collecting the eluant into a graduated cylinder instead of test tubes. We do this to avoid wasting time collecting fractions when we know that samples are not coming out yet. No sample can precede the void volume, so if we know the void volume is 30 mL for a column, there is no reason to collect fractions until we are close to that value. This means that you will have one volume collected in the graduated cylinder and more collected in the fractions. To calculate any elution volume, you must include all of the volume collected (the graduated cylinder and the fractions).

Second, it is often unclear what volume you would be recording for the eluting samples. Samples tend to spread out with any type of chromatography, so even if you applied only 0.5 mL of a protein solution, the protein would elute from the column in several milliliters. What volume would you call the elution volume? The V_0 is defined as the volume of the liquid outside the beads. If we load blue dextran on a Sephadex column to measure V_0, the actual volume of the column, which excludes the gel volume and the interior bead volume, would be the volume at which the blue dextran first started to come off. Some researchers prefer to plot absorbance vs. elution volume as in Figure 7.3 and then extrapolate the leading edge of the blue dextran peak back to where it crosses the x-axis and use that volume as the V_0. However, an accepted practice is to use the peak absorbance fraction to calculate the volume. This is easier because it relies on positive data for maximum absorbencies instead of trying to calculate where the first molecule of blue dextran came out. The elution volumes for the other components are calculated in the same way.

Column Calibration

There are different ways to determine the molecular weight of a molecule through gel filtration. If you are going to use a column repeatedly, the most accurate way is to take proteins of known molecular weight and run them as standards. This will give you a unique standard curve of log MW vs. V_e that can be used for that column and is definitely the safest way to go. If you wish to compare your data to those obtained from another experimenter, then you will need to have some type of relative elution volume. We have seen two ways of doing that in this chapter. One is to plot V_e/V_0. The other is to calculate a K_d or K_{avg}. To determine the latter, you need to know the total column volume, V_t, which can be determined in several ways. The easiest would be to just measure the liquid volume contained by the column at the height of the packed gel. You could also use an algebraic measurement based on the height and internal diameter. The elution volume of a molecule that has a K_d of 1 will also be nearly identical to V_t.

ESSENTIAL INFORMATION

Gel filtration is a powerful technique for purification and analysis of biomolecules. The most common gels are those produced by Pharmacia, which are called Sephadex, Sepharose, and Sephacryl. Each gel has a pore size that optimizes it for certain sizes of molecules. All molecules that are too big elute at the same time in the void volume. All molecules that are too small elute together in the total volume. The molecules that come out in between are in the fractionation range of the gel. These separate linearly according to the log of their molecular weights.

Experiment 7

Gel Filtration Chromatography

In this experiment, you will separate a three-component mixture using gel filtration chromatography on Sephadex G-75. One of the components is a protein, and you will also determine its molecular weight. This experiment can be done with or without fraction collectors and UV monitors depending on your available equipment.

Prelab Questions

1. What is the purpose of the slow loading procedure from steps 1–4? Why do we put small quantities of buffer on top of the column?

2. What must be done to the column fractions before they are assayed?

3. How will you know when it is time to stop running buffer through the column at the end of the experiment?

Objectives

1. Properly load a sample on the Sephadex gel without disturbing the gel bed.

2. Measure the peak elution volumes of blue dextran, DNP-aspartate, and a given protein.

3. Determine the K_d of the unknown protein.

4. Calculate the native molecular weight given the K_d you calculated and a table of known K_ds of standard proteins.

5. Use semilog paper to plot molecular weights.

Experimental Procedures

Materials

Sephadex G-75 (regular mesh), hydrated in 0.05 M TRIS-HCl, pH 7.5

Elution buffer, 0.1 M NaCl, 0.05 M TRIS-HCl, pH 7.5

Separation mixture, 3 mg blue dextran, 0.3 mg DNP-aspartate, 8 mg mystery protein, in 1 mL of elution buffer with 15% glycerol

Chromatographic column, 1.5 cm × 60 cm

Methods

1. Load 0.5 mL of the separation mixture onto the column. This part must be done *extremely carefully!* Your results will look terrible if you rush the loading process. The best way is to use a Pasteur pipet with a pipet pump. Slowly circle the pipet tip around the inside of the column near the resin to avoid making a pit in one spot of the column bed.

Tip 7.1

⟶ Beware of the loading process! Slow loading with a Pasteur pipet is the only way to do a decent job. Be very careful that you do not drop the pipet into the resin or that the pipet does not fall out of the pipet pump while you are loading.

---

2. Let the mixture run into the column while you are collecting the effluent into a graduated cylinder.

3. Do not add any more buffer to the top of the column until the mixture has completely entered. Then add small quantities and let the column run some more.

4. When the mixture is safely in the gel (2 cm from the top of the gel bed), you may fill the column to the top with elution buffer. Never let the top of the gel bed get completely dry.

5. When the blue dextran is getting close to exiting the column, remove the graduated cylinder and record the volume collected to this point. Begin to collect 1-mL fractions into cuvette-sized test tubes.

Analyzing Column Data Without UV Monitors

6. To establish the peak of elution for the blue dextran, add 1.5 mL of water to the fractions and assay with the spectrophotometers at 650 nm. This is the wavelength of maximum absorbance for blue dextran. Find the fraction with the highest absorbance.

7. Continue the elution and find the peak of elution of the unknown protein by measuring the absorbance of the tubes at 500 nm (after adding 1.5 mL of water). Although all proteins have a wavelength of maximum absorbance at 280 nm, this one happens to be reddish brown and can be detected with visible spectrophotometry at 500 nm.

8. Continue eluting the column and collecting the fractions until the DNP-aspartate has completely eluted and the column is clean. Find the peak elution fraction by measuring the absorbance at 440 nm (after adding 1.5 mL of water).

Analyzing Column Data with a Flow-Through UV Monitor

6. Although the three components all have different visible wavelengths of maximum absorbance, they can all be detected at 280 nm. Use the UV monitor and chart recorder to create a graph of absorbance at 280 nm vs. elution volume.

7. Determine the volumes that correspond to the three peaks.

Analysis of Results
Experiment 7: Gel Filtration Chromatography

Data

Turn in all of your raw data concerning the cumulative volume of buffer eluted from the column and the absorbance you recorded.

Calculations

1. On a single graph, plot the absorbance of the fractions versus the volume taken at that fraction. For example, if you collected 20 mL into the graduated cylinder and then began collecting the 1 mL fractions and your first dextran blue fraction that had an absorbance was on the fourth fraction, you would plot the absorbance versus 20 mL + 4 mL = 24 mL.

2. What were the elution volumes that gave the highest absorbencies for the blue dextran, the unknown protein, and the DNP-aspartate?

   Absorbance of blue dextran _____   cumulative volume _____ = V_0

   Absorbance of mystery protein _____   cumulative volume _____ = V_e

   Absorbance of DNP-aspartate _____   cumulative volume _____ = V_t

3. Calculate the K_d for the unknown protein.

$$K_d = \frac{V_e - V_0}{V_t - V_0} =$$

4. Use the information in Table 7.2 to make a graph of $\log_{10}$ MW vs. K_d. Now might be a good time to review the section on graphing in Chapter 1.

Table 7.2	K_ds for Some Known Proteins on Sephadex G-75	
Protein	**Molecular Weight**	K_d
Trypsin inhibitor (pancreas)	6,500	0.70
Trypsin inhibitor (lima bean)	9,000	0.60
Cytochrome-*c*	12,400	0.50
α-Lactalbumin	15,500	0.43
α-Chymotrypsin	22,500	0.32
Carbonic anhydrase	30,000	0.23
Ovalbumin	45,000	0.12

5. Interpolate from the graph and determine the MW of the unknown protein. This question does *not* ask you to determine its identity.

MW of mystery protein = _____

Questions

1. Why do large proteins come out of a gel filtration column faster than small ones? Wouldn't it make more sense for small things to move more quickly through a gel?

2. Would you be able to use Sephadex G-75 to separate alcohol dehydrogenase (MW 150,000) from β-amylase (MW 200,000)? Why or why not?

3. Would you be able to use Sephadex G-75 to separate alcohol dehydrogenase from bovine serum albumin? Why or why not?

Experiment 7a

Gel Filtration Chromatography of LDH

In this experiment, you will do the final column purification of LDH, using gel filtration chromatography with Sephadex G-150.

Prelab Questions

1. What is the purpose of the slow loading procedure? Why do we put small quantities of buffer on top of the column?

2. Where do you expect LDH to elute with Sephadex G-150? Why?

3. Why must we make sure the column does not run dry?

Objectives

1. Properly load a sample on the Sephadex gel without disturbing the gel bed.

2. Properly run a Sephadex gel without letting it run dry.

3. Locate and pool fractions containing LDH using a UV monitor and enzyme assay.

4. Estimate the native MW of LDH given a previously constructed graph of log MW vs. V_e/V_0.

Experimental Procedures

Preparation of Gel Filtration Columns

Materials

    1.5- $\times$ 60-cm Econo columns

    Sephadex G-150 (regular mesh) previously swelled in 0.05 M sodium phosphate, pH 7.5

Procedures

1. Acquire the columns and accessories you will need.

2. Acquire the column resin you will need.

3. Degas the Sephadex with a sidearm flask attached to a vacuum line until no more bubbles are coming off.

4. Remove fine particles of Sephadex in a graduated cylinder by siphoning off the last 5–10% of the volume of fuzzy particles.

5. Carefully pour the column with the Sephadex. Try very hard to avoid having the column completely pack before you have to add more. You want to get a smooth application of Sephadex to the system until it is tightly packed to near the top of the column.

6. Stopper the column well and wrap both top and bottom with Parafilm™. Store until needed.

7. It would be a good idea to check the column several hours before you need to use it just in case bubbles form, or it leaks and runs dry.

Gel Filtration Chromatography of LDH

Materials

1.5- × 60-cm Econo columns packed with Sephadex G-150

Fraction collectors

UV monitors

0.05 M Sodium phosphate, pH 7.5 (elution buffer)

Your purified LDH

Procedures

1. Acquire your Sephadex column and your concentrated LDH sample.

2. Assay your sample to be sure it has not lost activity during the storage process.

3. Assemble the column and peripheral hardware, setting the volume of the fraction collector to 1 mL.

4. Drain the excess buffer off the top of the gel, being very careful not to disturb the gel bed or let the resin run dry. This column is the most sensitive to the loading and running technique that we have used.

5. Carefully load your concentrated LDH onto the gel. From this point on, every drop counts as elution volume.

6. When the LDH sample is just entering the resin, slowly add 1 mL of elution buffer and let that drain in. Add another 1 mL and repeat this process until 5 mL have been added 1 mL at a time. Now you can *slowly* fill up the reservoir and keep the column running.

7. Assay your fractions and locate the ones containing enzyme, pooling those fractions that have significant activity. Assay the pooled fractions before proceeding to step 8.

8. Concentrate your samples down to 1 mL as in Experiment 6 and store at 4°C. This will be your final LDH sample, which we hope is very active and very pure.

9. As this was the last purification step, you must perform Experiment 3b, the protein assays, before completing the final purification table.

Analysis of Results

Experiment 7a: Gel Filtration of LDH

Data

1. Fill in the following table concerning the concentrated sample you loaded on the column.

Summary for Concentrated/Dialyzed Sample

Quantity assayed (μL)	
Dilution used (if any)	
Δ Absorbance/Δ min	
Units/mL	
Total mL loaded on column	
Total units loaded on column	

2. Turn in all of your raw data concerning the cumulative volume of buffer eluted from the column and the UV absorbance you recorded.

3. Record a void volume for the column. This may have been given to you, or you may have been asked to do the void volume yourself.

4. Fill in the following table for the elution fractions.

Data for Elution Fractions

Fraction #	Volume Assayed	$\Delta A/\Delta$ min	Units/mL	Total Units

Analysis of Results

Fill in the following table concerning your gel filtration results.

Summary of Gel Filtration Results

Fractions pooled	
Volume assayed	
Δ Absorbance/Δ min	
Units/mL	
Total units	
% Recovery off of column	
% Recovery from beginning of experiment	

Calculations

1. Calculate the V_e/V_0 ratio for LDH.

2. Using the graph shown in Figure 7.5, make a rough estimate of the LDH molecular weight. Is your answer reasonable?

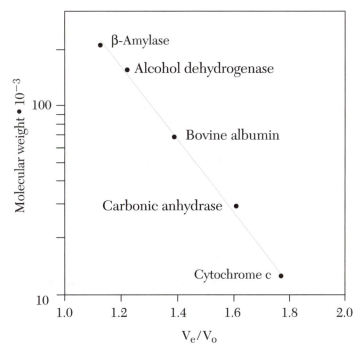

Figure 7.5   Log MW vs. V_e/V_0 for Sephadex G-150

3. Using the following table, finish your enzyme purification table.

Final Purification Table

Fraction	Units/mL	Total Units	% Recovery	mg/mL Protein	Specific Activity	Fold Purification
Crude						
20,000 × g super						
65% AS pellet						
Dialyzed 65% AS pellet						
Pooled IEX fractions						
Dialyzed IEX fractions						
Pooled Cibacron Blue fractions						
Concentrated Cibacron Blue fractions						
Pooled Sephadex fractions						
Concentrated Sephadex fractions						

Questions

1. What was the final % recovery of LDH that you saw? Is this reasonable?

2. What was the final fold purification that you saw? Is this reasonable?

3. Was there a particular place in the experiment where you lost too much activity?

4. What was the most effective step in the LDH purification?

5. What was the least effective step in the LDH purification?

6. If there are still contaminating proteins with your LDH, what would their physical characteristics be?

7. Give one example of a protein that might still be with your LDH.

8. Why do large proteins come out of a gel filtration column faster than small ones? Wouldn't it make more sense for small things to move more quickly through a gel?

9. Would you be able to use Sephadex G-75 to separate alcohol dehydrogenase (MW 150,000) from β-amylase (MW 200,000)? Why or why not?

10. Would you be able to use Sephadex G-75 to separate alcohol dehydrogenase from bovine serum albumin? Why or why not?

Additional Problem Set

1. If you chromatographed an oblong protein, why would it appear to be larger than a spherical protein of equal mass?

2. Using Table 7.1, explain how you would prepare an 80-mL column of an appropriate Sephadex gel needed to separate proteins ranging in size from bovine serum albumin to β-amylase.

3. Explain two advantages of using K_d values instead of just calculating the raw elution volume of your protein on your gel.

4. Explain how you could determine the exclusion limit of a gel of unknown origin.

5. What is the possible range of a K_d value? How does the value relate to the size of the protein?

6. Why is sodium azide often added to the storage buffer for a gel filtration column?

7. Why are gel filtration columns usually long and narrow whereas ion exchange columns are shorter and fatter?

Webconnections

For a list of Web sites related to the material covered in this chapter, go to **Webconnections** at the *Experiments in Biochemistry* site on the Saunders College Publishing Web page. You can access this page at http://www.saunderscollege.com. **Webconnections** are under *Experiments in Biochemistry* in the Biochemistry portion of the Chemistry page.

References and Further Reading

H. Ahern, (1996), *The Scientist* **10(5)**. Chromatography.

R. F. Boyer, *Modern Experimental Biochemistry*, Addison-Wesley, 1993.

A. Braithwaite and F. J. Smith, *Chromatographic Methods*, Blackie Academic and Professional, 1996.

M. K. Campbell, *Biochemistry*, Saunders College Publishing, 1998.

R. L. Dryer, and G. F. Lata, *Experimental Biochemistry*, Oxford University Press, 1989.

K. Robards, P. R. Haddad, and P. E. Jackson, *Principles and Practice of Modern Chromatographic Methods*, Academic Press, 1994.

J. F. Robyt and B. J. White, *Biochemical Techniques*, Waveland, Press, 1990.

T. Shihamoto, *Chromatographic Analysis of Environmental and Food Toxicants*, Marcel Dekker Publishing, 1998.

Sigma Chemical Company, *Biochemicals, Organic Compounds for Research, and Diagnostic Reagents*, 1995.

J. Stenesh, *Experimental Biochemistry*, Allyn and Bacon, 1984.

G. T. Tsao, *Chromatography*, Elsevier Science, 1991.

J. R. Whitaker, (1963), *Anal. Chem.* **35(12)**. Determination of Molecular Weights of Proteins by Gel Filtration on Sephadex™.

Chapter 8

Enzyme Kinetics

Topics

8.1 Reaction Rates

8.2 Order of Reactions

8.3 Michaelis–Menten Approach

8.4 Significance of K_m and V_{max}

8.5 Linear Plots

8.6 Properties of Tyrosinase

8.7 Why Is This Important?

Introduction

This chapter explores the nature of enzyme kinetics, which is the study of enzyme rates and their dependence on concentrations of enzyme, substrate, and the kinetic rate constants.

8.1   Reaction Rates

Enzyme kinetics is the study of enzyme rates and how these rates are affected by the concentration of the enzyme, the substrates, and any inhibitors or activators. The rate of the reaction is expressed as a change in the concentration of a reactant or product during a given time interval.

If a reaction can be written as

$$A + B \longrightarrow P$$

then the rate can be expressed in terms of either the appearance of the product, P, or the disappearance of either of the reactants, A or B. Thus

$$\text{Rate} = \Delta[P]/\Delta t = -\Delta[A]/\Delta t = -\Delta[B]/\Delta t$$

It has been established that the rate of the reaction at a given time is proportional to the product of the concentration of the reactants raised to an appropriate power,

$$\text{Rate } \alpha \text{ } [A]^a[B]^b \quad \text{or} \quad \text{Rate} = k[A]^a[B]^b$$

where k is a proportionality constant called a rate constant, and a and b are constants that must be determined experimentally.

8.2 Order of Reactions

The exponents in the rate equation are usually small whole numbers such as 0, 1, or 2. The values are dependent on the number of molecules of reactants and products involved and can often be deduced from the balanced chemical equation, but may also be more complicated to figure out.

If a reaction with the equation

$$A \longrightarrow B$$

can be shown to have the rate equation

$$\text{Rate} = k[A]^1$$

then it is said to be **first order** with respect to A. The rate at any time is governed by the rate constant times the concentration of A present at that time. A good example of this type of reaction is radioactive decay. The rate of radioactive decay is always based on the amount of radioactive substance present times a decay constant.

If the rate of a reaction

$$A + B \longrightarrow C + D$$

is governed by the rate equation

$$\text{Rate} = k[A]^1[B]^1$$

then the reaction is *first order* with respect to A and *first order* with respect to B, but *second order* overall. This rate would go faster if you increased the concentrations of either A or B or both.

8.3 Michaelis–Menten Approach

The model for enzyme kinetics most often seen was proposed in 1913 by Michaelis and Menten. Although it has undergone much revision, it is still the basic one used for all non-allosteric enzymes.

An enzyme-catalyzed reaction can be written

$$E + S \underset{k_2}{\overset{k_1}{\rightleftharpoons}} ES \underset{k_4}{\overset{k_3}{\rightleftharpoons}} E + P$$

where E = free enzyme
S = free substrate
ES = enzyme–substrate complex
P = product
k_n = individual rate constants

Usually we perform these experiments for short periods so that there is no significant buildup of product, P. Therefore, there is no back-reaction, k_4. The equation then simplifies to

$$E + S \underset{k_2}{\overset{k_1}{\rightleftharpoons}} ES \overset{k_{cat}}{\longrightarrow} E + P$$

where k_{cat} is the catalytic rate constant for breakdown to product.

The rate of an enzyme-catalyzed reaction is dependent on the concentrations of enzyme and of substrate, but the relationships are not the same for both. If we plot rate, v, vs. [E], we get a curve as in Figure 8.1. This is a straight line. This has important implications for our ability to do an enzyme kinetic experiment. Any change in the amount of enzyme will change the rate that we observe, so we must pipet our enzyme very carefully.

However, if we make a similar plot of v vs. [S], we get a different graph (see Figure 8.2). The rate is dependent on the amount of substrate present, but the line is not straight. At

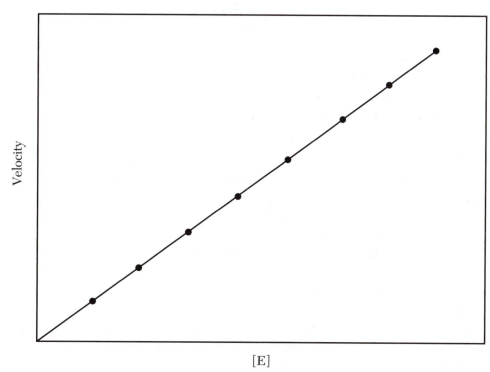

Figure 8.1 Reaction velocity vs. enzyme concentration

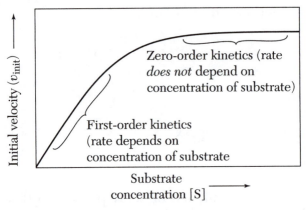

Initial velocity (v_{init}) →

Zero-order kinetics (rate *does not* depend on concentration of substrate)

First-order kinetics (rate depends on concentration of substrate

Substrate concentration [S] →

Figure 8.2 Velocity vs. substrate for the Michaelis–Menten model

low substrate concentrations, the line is straight and the rate is first order with respect to S. Then the rate flattens out and eventually reaches zero order with respect to S. This means that at high [S], the rate is independent of the amount of S present. This is a hyperbolic curve, with an asymptote at what we call V_{max}. V_{max} is the theoretical maximum velocity, which occurs at infinitely high [S]. *At high [S], the velocity increases less with each increase of S.*

To account for this observation of kinetic properties, the Michaelis–Menten equation was derived:

$$v = \frac{V_{max}[S]}{K_m + [S]}$$

where v = **initial velocity**
V_{max} = **maximum velocity**
[S] = **substrate concentration**
K_m = **Michaelis constant** = $(k_2 + k_{cat})/k_1$

The v is the initial velocity and must be measured early in the reaction, before product has built up. The V_{max} is the velocity when all of the enzyme active sites are filled with substrate. This is theoretical because it would never really happen. There would always be some free E around. The Michaelis constant, K_m, is numerically equal to the substrate concentration that gives half the maximal velocity. This can be estimated from a Michaelis–Menten graph (Figure 8.3). By estimating the V_{max}, you can calculate what half of V_{max} would be, go over to the curve, and drop down to the concentration of substrate. Where the [S] gives $V_{max}/2$ is the K_m.

V_{max} and K_m are usually referred to as constants and are the first two quantities determined for an enzyme-catalyzed reaction. It should be pointed out that the V_{max} is not a constant at all. It depends on the total [E] and is therefore only a constant for your particular experiment. K_m is a constant, however, at least under given conditions of pH, ionic strength, and choice of substrate. If we increase the [E], we will increase the velocity at any [S], but the [S] that gives $V_{max}/2$ will not change.

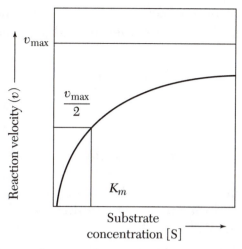

Figure 8.3 Determination of K_m and V_{max} with a Michaelis–Menten graph

8.4 Significance of K_m and V_{max}

K_m is important for several reasons. We have said that it tells the [S] needed to reach $V_{max}/2$. That in itself may not seem terribly important, but it allows us to better plan our experiments. If you want to test the effect of [S] on v, then you will choose a concentration of S that is close to K_m, so that a small change will affect the rate (see Figures 8.2 or 8.3). However, if you want to measure the quantity of enzyme in a system, then you want the [E] to be the only variable affecting the rate. Therefore, you choose a [S] that is very high so that there is no effect of [S] on v. For such a test you would use [S] of 5 to 10 times the K_m.

K_m is also important because it gives an idea of the affinity between the enzyme and substrate. $K_m = (k_2 + k_{cat})/k_1$, or the rate of breakdown of ES divided by the rate of formation of ES. Often, the catalytic rate constant, k_{cat}, is much slower than the individual rate constants for formation and breakdown, k_1 and k_2. Under those circumstances, K_m is the same as K_s, a true dissociation constant, and a low K_m means that the ES complex forms rapidly. Enzymes often have multiple substrates. A substrate with a lower K_m will bind more quickly to the enzyme than one with a high K_m.

If you isolated a new enzyme from human liver and determined that the K_m for the substrate was 5 mM, but the naturally occurring level of that substrate in the liver was 5 nM, then you would be able to conclude one of the following: (a) The enzyme was damaged somehow, (b) the reaction does not naturally happen quickly, or (c) the substrate you isolated is not the true or best substrate.

In a clinical lab, blood or other samples are often screened for various enzymes. Sometimes it is the total level of enzyme that is measured, where an increase or decrease may indicate a disease state. Other times it is the K_m that is measured because an altered K_m may indicate a damaged enzyme or another form of the enzyme that should not be there.

V_{max} is also important, although it is not a constant itself. The basic rate law is as follows:

$$v = k_{cat}[ES]$$

The constant, k_{cat}, is the catalytic rate constant, and the enzyme and substrate must be in the ES form to react. As we said earlier, the theoretical V_{max} is achieved if all of the enzyme is in the ES form, so

$$V_{max} = k_{cat}[E_t]$$

where E_t = total concentration of enzyme.
    Rearranging gives

$$k_{cat} = V_{max}/[E_t]$$

By calculating the V_{max} and the actual molar concentration of enzyme, we can calculate k_{cat}. This constant is also called the **turnover number,** which is the number of substrate molecules transformed to product per unit time by one enzyme molecule under maximal conditions. This is a measure of the speed and efficiency of an enzyme.

8.5   Linear Plots

We could estimate K_m and V_{max} from a curve similar to Figure 8.2, but there are problems with this. First, because V_{max} is never really attained, we can't really measure it. Second, because we plot K_m as $V_{max}/2$, it too has error. Luckily, many investigators have manipulated the Michaelis–Menten equation to give a linear plot. The most common is the well-known Lineweaver–Burk plot shown in Figure 8.4. If we rearrange the Michaelis–Menten equation and take reciprocals, we get:

$$1/v = K_m/V_{max}[S] + 1/V_{max}$$

where $1/V_{max}$ is the y-intercept and $-1/K_m$ is the x-intercept.

◉▸ Practice Session 8.1

Using the data given, calculate the K_m and V_{max}. Calculate the turnover number assuming you used 0.1 mL of a 0.3 mg/mL solution of enzyme with a molecular weight of 136,000.

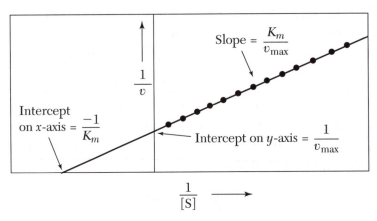

Figure 8.4   Lineweaver–Burk plot

Velocity vs. [S] Data

Velocity (μmol/min)	[S] (M)
130	6.5×10^{-4}
116	2.3×10^{-4}
87	7.9×10^{-5}
63	3.9×10^{-5}
30	1.3×10^{-5}
10	3.7×10^{-6}

First, change the data to reciprocals so that you can use the Lineweaver–Burk plot.

Reciprocal Data

$1/v$ (min/μmol)	$1/[S]$ (L/mol)
0.00769	1.539×10^{3}
0.00862	4.35×10^{3}
0.0115	1.27×10^{4}
0.0159	2.56×10^{4}
0.0333	7.69×10^{4}
0.1000	2.7×10^{5}

Then plot the values on a graph as in Figure 8.5. From the graph, the y-axis $= 1/V_{max} = 0.006$.

$$V_{max} = 1/0.006 = 166.6\ \mu\text{mol/min}$$

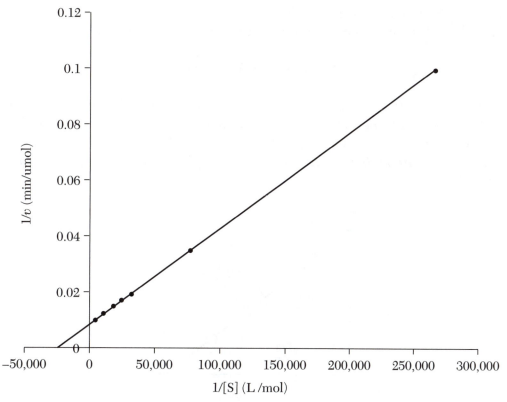

Figure 8.5 Lineweaver–Burk plot

From the x-axis, $-1/K_m = 117.5 \times 10^3$ M^{-1}.

$$K_m = 5.7 \times 10^{-5} \text{ M}$$

To calculate the turnover number, you need to figure out how many moles of enzyme you used. The molecular weight of the enzyme was given as 136,000. You used 0.1 mL of a 0.03 mg/mL solution.

$$0.3 \text{ mg/mL} \times 0.1 \text{ mL} = 0.03 \text{ mg of enzyme}$$

$$0.03 \text{ mg}/136{,}000 \text{ mg/mmol} = 2.2 \times 10^{-7} \text{ mmol} = 2.2 \times 10^{-4} \text{ } \mu\text{mol}$$

$$\text{Turnover number} = k_{cat} = V_{max}/\text{enzyme} = 166.6 \text{ } \mu\text{mol/min} \div 2.2 \times 10^{-4} \text{ } \mu\text{mol}$$

$$= 757{,}273/\text{min}$$

Alternative Plots

For the information of those of you who have done the Lineweaver–Burk plot before, or those who recognize its inherent inferiority, many other plots are more popular nowadays. The Eadie–Hofstee plot uses

$$v = -K_m(v/[S]) + V_{max}$$

Velocity is plotted versus $v/[S]$. The y-intercept gives the V_{max}; the x-axis gives the V_{max}/K_m, and the slope is $-K_m$. This plot is considered superior to the Lineweaver–Burk plot because it has only one reciprocal, which reduces errors, and the substrate concentrations are weighted equally. In addition, the points do not cluster around the y-axis. A picture of an Eadie–Hofstee plot is shown in Figure 8.6.

The Eisenthal–Cornish–Bowden plot uses the following equation:

$$V_{max} = v + K_m(v/[S])$$

With this plot the points are not plotted as on a normal graph. The coordinates for v and $[S]$ are plotted directly on their axes and then the points are connected. Where the lines intersect gives V_{max} on the y-axis and K_m on the x-axis as in Figure 8.7. *This is the easiest and one of the most accurate plots for quickly determining K_m and V_{max} if you are drawing the graph by hand.*

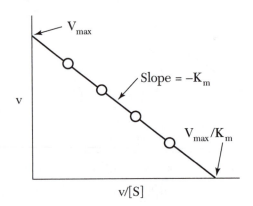

Figure 8.6 Eadie–Hofstee Plot

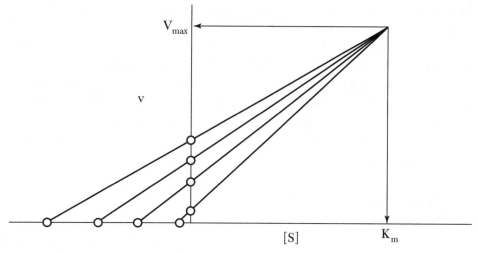

Figure 8.7 Eisenthal–Cornish–Bowden Plot

8.6   Properties of Tyrosinase

Tyrosinase, also called polyphenol oxidase, is present in plant and animal cells. In mammalian cells, it catalyzes two steps in the synthesis of melanin pigments from tyrosine. It is located in the skin and is activated by UV light from the sun, leading ultimately to a suntan.

Mushroom tyrosinase is tetrameric, with a total molecular weight of 128,000. Four Cu^+ atoms are associated with the active enzyme. One of the natural substrates of tyrosinase is 3,4-dihydroxyphenylalanine (DOPA). To study the kinetics of tyrosinase, we will use the reaction shown in Figure 8.8 and the fact that the product, dopachrome, absorbs at 475 nm.

8.7   Why Is This Important?

To understand the nature of metabolism, we must understand the nature of the enzymes that catalyze the reactions. Enzyme kinetics is the study of reaction rates of enzyme-catalyzed reactions. By observing how reaction rates are affected by the concentration of enzyme, substrate, inhibitors, and activators, we will begin to understand the nature of the reaction. For example, if we look at the K_ms for two enzymes that catalyze the formation of

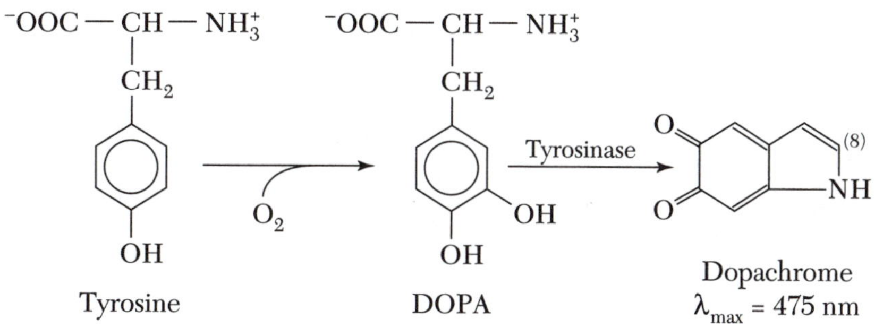

Figure 8.8 The tyrosinase reaction

glucose 6-phosphate from glucose, hexokinase, and glucokinase, we can tell something about the importance of these two enzymes. The K_m for glucose with hexokinase is 0.15 mM and for glucose with glucokinase is 20 mM. Glucokinase occurs only in the liver, but almost all cells have hexokinase. The normal level of blood glucose is 5 mM, so we could predict that hexokinase will be utilizing glucose well in all tissues. Glucokinase, on the other hand, would only process glucose when the blood glucose was higher, such as after a carbohydrate-rich meal. This would mean that when glucose is low, the other tissue would have a better opportunity to use it. Only when glucose is high will the liver use its other enzyme, glucokinase, to help process it.

ESSENTIAL INFORMATION

The kinetics of an enzyme-catalyzed reaction can tell us a lot about the nature of the enzyme and the metabolism involved. It was discovered long ago that the relationship of the reaction rate to the amount of enzyme was different from the relationship of the reaction rate to the amount of substrate. This difference led to our current model of simple enzyme-catalyzed reactions. The reaction rate is linearly dependent on the amount of enzyme. Therefore, you must always be careful when you pipet the enzyme. For non-allosteric enzymes, the rate is hyperbolic to the substrate concentration. The rate increases quickly with increasing substrate at low substrate levels, but does not increase much at all with the same level of increase if the substrate level is already high. When one wants to compare amounts of enzymes between different samples, one uses high concentrations of substrate. In this way, the only variable is the enzyme itself. When one wants to study the effect of substrate on enzymes, one uses a broad range of substrate levels, from low to very high. Two parameters that are often calculated are the K_m and the V_{max}. The V_{max} is the maximum rate you would see with infinite substrate concentration. The K_m is the concentration of substrate that gives half the rate of the V_{max}. Many types of plots are available to determine these parameters. The most common linear plot is the Lineweaver–Burk plot. The simplest and fastest is the Eisenthal–Cornish–Bowden plot.

Experiment 8

Enzyme Kinetics of Tyrosinase

In this experiment, you will study enzyme kinetics using mushroom tyrosinase. Enzyme and substrate will be varied to demonstrate the different nature of the effects of each on reaction velocity. The kinetic parameters K_m, V_{max}, and k_{cat} will be determined.

Prelab Questions

1. For the first part of the experiment, where you are varying the amount of tyrosinase, how many mL of 15 mM L-DOPA will be in each tube?

> **Tip 8.1**
>
> ⮑ The beauty of this type of experiment is that each tube is its own reagent blank, because you are changing absorbencies. Therefore, you don't have to limit yourself to just one cuvette.

2. If I gave you two tubes and told you that one was enzyme and the other was substrate, what experiment could you do to figure out which was which? For this, you have no other source of enzyme or substrate to use, but you do have spec tubes and a spectrophotometer.

Objectives

Upon successful completion of this unit, the student will be able to

1. Observe and analyze the effects of changing enzyme concentration on initial reaction rates.

2. Observe and analyze the effect of changing substrate concentration on initial reaction rates.

3. Construct v vs. [S] plots and estimate K_m and V_{max}.

4. Construct Lineweaver–Burk plots and determine K_m and V_{max}.

5. Calculate k_{cat} given total enzyme concentrations.

6. Design protocols for determining kinetic parameters.

7. Explain the significance of K_m, V_{max}, and k_{cat}.

Experimental Procedures

Materials

Sodium phosphate buffer, 0.1 M, pH 7.0

Mushroom tyrosinase (0.2 mg/mL)

L-DOPA, 15 mM

Spectrophotometers

Cuvettes

Methods

Part A: Tyrosinase Level for Kinetic Assays

In this part of the experiment, you will determine the proper amount of tyrosinase to use for the kinetic analyses. If too little enzyme is used, then the change in A_{475} will be too small to detect, especially at low substrate. On the other hand, if too much is used, the substrate will be depleted too quickly and the rate will not be linear for a measurable time. For the dopachrome assay, you want a linear rate for at least a minute while taking time points at 15-s intervals.

1. Set up a protocol for the determination of the correct tyrosinase concentration. Each tube should have a total of 3 mL of solution and be 5 mM in L-DOPA. The volume will be controlled by the phosphate buffer, and the amount of enzyme will vary. Do at least five tubes ranging from 0.1 to 0.5 mL of tyrosinase.

2. Pipet the L-DOPA and phosphate buffers.

Tip 8.2

⊃ *Never* vortex enzyme solutions! Always mix by quick but gentle inversion using Parafilm to cover the top of the tube.

---

3. Pipet the chosen amount of enzyme into one tube. Mix by inversion and read the change in absorbance at 475 nm immediately.

4. Measure for 2 min, recording at 15-s intervals. Repeat steps 2–4 for the other enzyme concentrations. *Remember to add the enzyme immediately before reading each tube!*

5. Determine the rate for five different tyrosinase levels. The final amount chosen for the next part should give you a change in absorbance per minute ($\Delta A/\Delta$ min) between 0.2 and 0.3/min. The levels given in step 1 are only suggestions. You may have to try higher or lower concentrations.

Part B: Kinetic Constants of Tyrosinase

Now that the appropriate enzyme level has been determined, the kinetic parameters K_m, V_{max}, and k_{cat} will be determined. In the first part, the L-DOPA was saturating. In this part, nonsaturating levels will be used.

1. Set up a protocol as before using the level of tyrosinase that you chose in the first part. There should still be 3 mL per tube. The recommended levels of L-DOPA are 0.05, 0.10, 0.20, 0.40, 0.80, and 1.0 mL.

2. Follow the same procedures for addition, mixing, and recording as in the last part.

3. Be sure that a rate is attained that is close to V_{max}. That is, make sure that 1.0 mL of L-DOPA is close to saturating.

Analysis of Results

Experiment 8: Enzyme Kinetics of Tyrosinase

Data

Part A

Indicate the absorbencies at the time points you used for the varying enzyme volumes.

Enzyme Volume (mL)	Start	15″	30″	45″	60″	75″	90″	105″
0.1								
0.2								
0.3								
0.4								
0.5								

Part B

Indicate the absorbencies recorded for the time points you used for the varying substrate levels.

Substrate Volume (mL)	Start	15″	30″	45″	60″	75″	90″	105″
0.05								
0.10								
0.2								
0.4								
0.8								
1.0								

Calculations

The concentration of the tyrosinase is 0.2 mg/mL. The extinction coefficient for dopachrome at 475 nm is $3600 \text{ M}^{-1} \text{ cm}^{-1}$.

Part A

1. Calculate the rate of reactions as $\Delta A / \Delta$ min. You may want to graph the absorbencies vs. time to establish what the initial velocity is. If so, turn in your graphs with this write-up. If you choose to calculate it otherwise, show how you calculated the rate.

2. Plot rate (μmol/min) vs. amount of E in mg. The changes in absorbance per minute must be converted to change in μmol using Beer's law and the extinction coefficient for dopachrome. This is similar to the calculations we did for LDH earlier:

$$\mu\text{mol product/min} = \frac{(\Delta A / \Delta \text{ min})}{3600 \text{ M}^{-1}} \times (10^6 \ \mu\text{M/M}) \times 0.003 \text{ L}$$

Enzyme Volume (mL)	Δ A$_{475}$/min	μmol/min
0.1		
0.2		
0.3		
0.4		
0.5		

What amount of enzyme did you choose to use in part B?

Part B

1. Calculate the number of units of enzyme activity for the varying amounts of substrate. Calculate the substrate concentration in the test tubes at the time = zero point.

Substrate Volume (mL)	L-DOPA (mM)	Δ A$_{475}$/min	μmol/min
0.05			
0.1			
0.2			
0.4			
0.8			
1.0			

2. Make a Michaelis–Menten graph. Plot v (μmol/min) vs. [S] (mM). Estimate K_m and V_{max} from this graph.

K_m from Michaelis–Menten Graph _____

V_{max} from Michaelis–Menten Graph _____

3. Make any *linear* plot (L-B, ECB, EH, etc.) to determine the K_m and V_{max}.

   K_m from linear plot _____

   V_{max} from linear plot _____

4. Use the molecular weight of tyrosinase, 128,000, to determine how many μmol of tyrosinase are in the tubes that you used to calculate V_{max}.

5. The turnover number is the number of moles of product produced per minute by 1 mol of enzyme at V_{max}. Using the number you calculated in step 4 and the V_{max} from your linear graph, calculate the turnover number for tyrosinase.

Questions

1. A competitive inhibitor competes with substrate for binding to the active site of the enzyme. The enzyme, once bound by the inhibitor, is unable to form product. How would a competitive inhibitor affect the velocity of product formation? Would you need more or less of the substrate to get the same velocity as found before the inhibitor was added?

2. Assume you have an enzyme that catalyzes a reaction that breaks down dopachrome. At $t = 0$, the absorbance at 475 is 0.2 when you add the enzyme. At $t = 30$ s, would you expect the absorbance to be less than or greater than 0.2?

3. Enzyme X has a molecular weight of 48,000. It converts substrate Z into product Y. Z absorbs at 340 nm, and Y absorbs at 480 nm.
   a. At what wavelength would you measure the change in absorbance to assay for enzyme X? Would the absorbance increase or decrease over time?
   b. If V_{max} = 60 μmol/min and you used 400 μL of a 0.1 mg/mL solution of enzyme, what is the turnover number?

Experiment 8a

Enzyme Kinetics of LDH

In this experiment, you will study the kinetic parameters of your lactate dehydrogenase (LDH). You will use linear plots to determine the K_m for NAD^+ and the V_{max}. You will then calculate an estimated turnover number, k_{cat}.

Objectives

Upon successful completion of this unit, the student will be able to

1. Observe and analyze the effects on initial reaction rates of changing enzyme concentration.

> **Tip 8.1**
> ⟶ The beauty of this type of experiment is that each tube is its own reagent blank, because you are changing absorbencies. Therefore, you don't have to limit yourself to just one cuvette.

2. Observe and analyze the effects on initial reaction rates of changing substrate concentration.
3. Construct v vs. [S] plots and estimate K_m and V_{max}.
4. Construct Lineweaver–Burk plots and determine K_m and V_{max}.
5. Calculate k_{cat} given total enzyme concentrations.
6. Design protocols for determining kinetic parameters.
7. Explain the significance of K_m, V_{max}, and k_{cat}.

Experimental Procedures

Materials

CAPS buffer, 0.14 M, pH 10

NAD^+, 6 mM

Lactate, 0.15 M

Spectrophotometers

Cuvettes

Procedures

1. Using your best sample of LDH and the standard assay, vary the amount of LDH and calculate the initial velocity for five different volumes. Find an amount that gives a change in absorbance vs. time of about 0.2/min using the standard assay. You may have to experiment with this to get it. The acceptable range would be 0.15 to 0.25.

2. Once you find the correct amount of LDH to use, you will hold this value constant for the duration of this experiment.

3. Using the standard NAD^+ solution, adjust the assay so that you add progressively less of the NAD^+ each time. Add water to compensate for the lost NAD^+ volume.

4. Record the absorbance changes over time and calculate the $\Delta A/\Delta$ min. If you get down to 50 μL of the NAD^+ and you still are not seeing much decrease in activity, make a dilution of the NAD^+ and try again.

5. When you have assayed volumes of NAD^+ that run a range from full activity down to almost no activity and this includes at least six good assays, you probably have enough data to analyze.

6. Calculate the concentration of NAD^+ in each cuvette at the beginning of the assay and the initial activity recorded in μmol/min.

7. Use your favorite linear graph to calculate the K_m for NAD^+ and the V_{max} for the amount of enzyme you chose.

8. Using the known protein concentration of the LDH sample you used, and assuming (falsely) that all of the protein is LDH, calculate the turnover number or k_{cat} for LDH assuming a molecular weight of 150,000.

Name _____ Section _____

Lab partner(s) _____ Date _____

Analysis of Results

Experiment 8a: Enzyme Kinetics of LDH

Data

Indicate the absorbencies at the time points you used for the varying enzyme volumes.

Enzyme Volume (mL)	Start	15″	30″	45″	60″	75″	90″	105″

Indicate the absorbencies recorded for the time points you used for the varying substrate levels.

Substrate Volume (mL)	Start	15"	30"	45"	60"	75"	90"	105"

Calculations

1. Calculate the rate of reactions as $\Delta A/\Delta$ min. You may want to graph the absorbencies vs. time to establish what the initial velocity is. If so, turn in your graphs with this write-up. If you choose to calculate it otherwise, show how you calculated the rate.

2. Plot rate (μmol/min) vs. amount of E in mL. The changes in absorbance per minute must be converted to change in μmol using Beer's law and the extinction coefficient for NADH.

$$\mu\text{mol product/min} = \frac{(\Delta A/\Delta \text{ min})}{6220 \text{ M}^{-1}} \times \frac{10^6 \ \mu\text{M}}{\text{M}} \times 0.003 \text{ L}$$

Enzyme Volume (mL)	Δ A$_{340}$/min	μmol/min

What is the amount of enzyme that you chose to use for the second part with varying substrate?

3. Calculate the number of units of enzyme activity for the varying amounts of substrate. Calculate the substrate concentrations in the test tubes at the time = zero point.

Substrate Volume (mL)	mM NAD$^+$	Δ A$_{340}$/min	μmol/min

4. Make a Michaelis–Menten graph. Plot v (μmol/min) vs. [S] (mM). Estimate K_m and V_{max} from this graph.

   K_m from Michaelis–Menten graph _____

   V_{max} from Michaelis–Menten graph _____

5. Make any *linear* plot (L-B, ECB, EH, etc.) to determine the K_m and V_{max}.

   K_m from linear plot _____

   V_{max} from linear plot _____

6. Use the molecular weight of LDH, 150,000, and the protein concentration you calculated previously to determine how many micromoles of LDH are in the tubes that you used to calculate V_{max}.

7. The turnover number is the number of moles of product produced per minute by one mole of enzyme at V_{max}. Using the number you calculated in step 6 and the V_{max} from your linear graph, calculate the turnover number for LDH.

8. If you wanted to determine the K_m for lactate, what protocol would you set up?

9. Why would the clever lab instructor choose to have you determine the K_m for NAD^+ instead of for lactate?

10. Which of the components of the assay for K_m of NAD^+ would have the least effect on your results if they were pipetted imprecisely? Why?

Additional Problem Set

1. For the hypothetical reaction

$$3A + 2B \longrightarrow 2C + 3D$$

the rate was experimentally determined to be

$$\text{rate} = k[A]^1[B]^1$$

What is the order of the reaction with respect to A? With respect to B? What is the overall order of the reaction?

2. For an enzyme that displays Michaelis–Menten kinetics, what is the reaction velocity, v (as a percentage of V_{max}), observed at (a) $[S] = K_m$; (b) $[S] = 0.5\ K_m$; (c) $[S] = 0.1\ K_m$; (d) $[S] = 2\ K_m$; (e) $[S] = 10\ K_m$?

3. How is the turnover number related to V_{max}?

4. Why is V_{max} not a true constant?

5. Determine the K_m and V_{max} for an enzymatic reaction, given the data shown in the following table.

Substrate Concentration (mM)	Velocity (μmol/min)
2.500	0.588
1.000	0.500
0.714	0.417
0.526	0.370
0.250	0.256

6. Why do we do linear transformations of the Michaelis–Menten equation?

7. A standard line fits the equation $y = mx + b$, where m is the slope and b is the y-intercept. What are these values for the equations for (a) Lineweaver-Burk plot, (b) Eadie–Hofstee plot, and (c) Eisenthal–Cornish–Bowden plot?

Webconnections

For a list of Web sites related to the material covered in this chapter, go to **Webconnections** at the *Experiments in Biochemistry* site on the Saunders College Publishing Web page. You can access this page at http://www.saunderscollege.com. **Webconnections** are under *Experiments in Biochemistry* in the Biochemistry portion of the Chemistry page.

References and Further Reading

R. F. Boyer, *Modern Experimental Biochemistry*, Addison-Wesley, 1993.
M. Campbell, *Biochemistry*, Saunders College Publishing, 1998.
A. Cornish-Bowden, *Analysis of Enzyme Kinetic Data*, Oxford University Press, 1995.
R. L. Dryer and G. F. Lata, *Experimental Biochemistry*, Oxford University Press, 1989.

H. Gutfreund, *Kinetics for Life Sciences: Receptors, Transmitters, and Catalysts,* Cambridge University Press, 1995.

R. C. Jack, *Basic Biochemical Laboratory Procedures and Computing,* Oxford University Press, 1995.

R. J. Leatherbarrow, (1990), *Trends Biol. Sci.* **15.** Linear and Non-Linear Regression of Biochemical Data.

D. L. Purich, J. N. Abelson, and M. I. Simon, *Contemporary Enzyme Kinetics and Mechanisms,* Academic Press, 1996.

J. F. Robyt and B. J. White, *Biochemical Techniques,* Waveland Press, 1990.

I. H. Segel, *Biochemical Calculations,* 2nd ed., Wiley Interscience, 1976.

A. R. Schulz, *Enzyme Kinetics: From Diastase to Multi-Enzyme Systems,* Cambridge University Press, 1994.

Chapter 9

Electrophoresis

Topics

Introduction

This chapter introduces what is probably the most important biochemistry technique for you to learn. Electrophoresis is used to separate biological molecules in an electric field. It is most often used to separate proteins or DNA, and it can be done with a system based on agarose or polyacrylamide. All of the chapters and experiments that follow rely heavily on electrophoresis.

9.1   Electrophoresis

Electrophoresis is the movement of charged particles in an electric field. A negatively charged particle will move toward the positive pole and vice versa. The velocity at which it moves depends on several factors according to the following equation:

$$v = \frac{qE}{f}$$

where v = **velocity**

q = **net charge on the molecule**

E = **applied voltage**

f = **frictional coefficient**

Therefore, a molecule with a charge of −2 will move twice as fast, all else being equal, as a molecule with a charge of −1. Remember that the net charge of most biological molecules depends on the medium in which you put them. If you change the pH of the buffer in the system, you will change the net charges on some of the molecules.

If you increase the voltage, the separation will also be quicker, but there are limitations on this. Too high a voltage can destroy your sample or the support medium due to excessive heat. Also, the bands are usually sharper and better separated with less streaking if the electrophoresis is done slowly. In a teaching lab, we have time limitations, so sometimes we put up with less than perfect results so that you get out at a reasonable hour.

The frictional coefficient is the retarding effect of the size and shape of the particle and the nature of the support medium. A round protein, for example, will move faster than a rod-shaped one with the same weight. A denser medium retards all proteins but has a greater effect on larger ones.

Although electrophoresis could be used to separate charged molecules from any class of biomolecule, the two most common are proteins and nucleic acids. This chapter will only discuss separation of proteins because Chapter 11 deals with DNA separation.

Proteins are differentiated from one another based on their amino acid sequence. The amino acid sequence gives each protein a unique charge character as well as size and shape. All of these factors act together to affect how the proteins migrate with electrophoresis.

Many different media can be used for electrophoresis, such as liquid, paper, or gels. Most electrophoresis done today uses a gel-based medium.

9.2 Agarose Gels

Agarose is one of the most common supports for electrophoresis. Agarose is a natural polysaccharide of galactose and 3,6-anhydrogalactose derived from agar, which is itself obtained from certain red algae. The repeating unit is shown in Figure 9.1.

Agarose chains tend to make left-handed helices that intertwine with each other. This gives rise to a gel that is quite dense for its concentration. A solution of only 1.0% w/v agarose

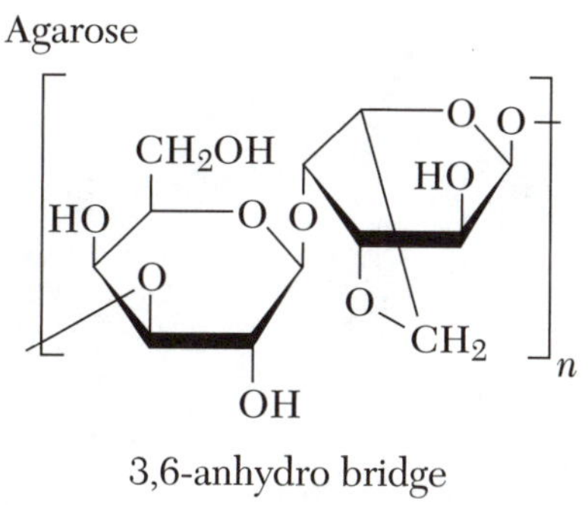

Agarose

3,6-anhydro bridge

Figure 9.1 The structure of agarose

will solidify into a fairly dense gel that can be used to separate proteins or nucleic acids. The agarose gels are prepared by boiling a defined quantity of the dry polymer in buffer until it melts. The melted agarose suspension is then poured into a casting tray with a well-forming comb and allowed to cool and solidify. This type of gel makes an excellent support for separating proteins based on charge. The spongelike, porous gel will allow proteins to pass through quickly but provides enough stability so that they do not diffuse quickly when the power is turned off. It is important to remember that the ions in the buffer carry most of the current during the electrophoresis. This buffer is also in the gel, so some of the current goes through the gel so the samples can move. If you accidentally made your gel up in water, there would be no current flowing through the gel and your samples would not move.

Tip 9.2

⮑ Weigh out your agarose carefully because there is a huge difference between a 0.8% gel and a 1% gel. Be sure not to let your agarose boil too long or it will become too concentrated.

Agarose gels are usually run horizontally in the submarine mode, where the gel is completely immersed in buffer. This aids in heat reduction. Figure 9.2 shows an agarose gel setup. Agarose gels are **native gels,** because nothing in the system denatures a protein if

ESSENTIAL INFORMATION

Agarose gels are native gels, in which the molecule of interest retains its native conformation and activity while it runs. Agarose is simple to use. You just have to heat the correct amount in a suitable buffer until the agarose melts and then pour your gel. Small differences in concentration are very important. A 0.7% agarose gel is very flimsy; a 1.2% gel is a brick. Always make up agarose gels in a buffer, not water. When proteins separate on an agarose gel, the size, shape, and charge of the protein determine how far it travels in the allotted time. Running the gel at higher voltage will speed up the process but reduce the quality of your results.

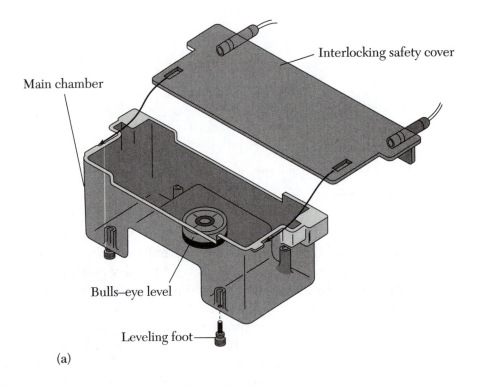

Interlocking safety cover

Main chamber

Bulls–eye level

Leveling foot

(a)

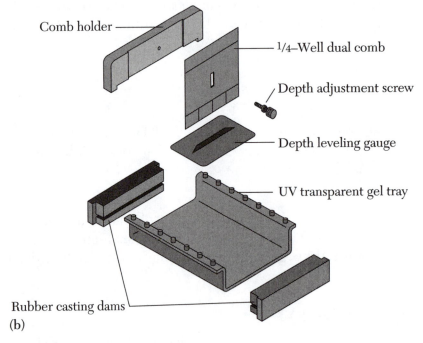

Comb holder

1/4–Well dual comb

Depth adjustment screw

Depth leveling gauge

UV transparent gel tray

Rubber casting dams

(b)

Figure 9.2 Submarine gel setup

everything goes according to plan. Proteins retain their normal conformations and activity, if any, as they run down the gel. However, if you have an active enzyme, you must be careful that the temperature does not get too high during the run; otherwise you may denature your enzyme enough to destroy its biological activity. Running time, voltage, and buffer volume all control the heat output of the electrophoresis.

9.3 Polyacrylamide Gels

Polyacrylamide gels are long polymers of acrylamide cross-linked with N,N'-methylene-bisacrylamide, as shown in Figure 9.3. Unlike the simple agarose gels, there are many components to a polyacrylamide gel, and the nature of the gel is controlled by the amounts chosen.

Total Concentration of Acrylamide

The mobility, μ, for an electrophoresis is the ratio of the velocity of the particle to the voltage you applied. The following equation describes the relationship of mobility to acrylamide concentration:

$$\mu = e^{-kT}$$

where k is a constant for a certain percentage of bisacrylamide and T is the total acrylamide concentration. Therefore, the higher the concentration, the slower particles move.

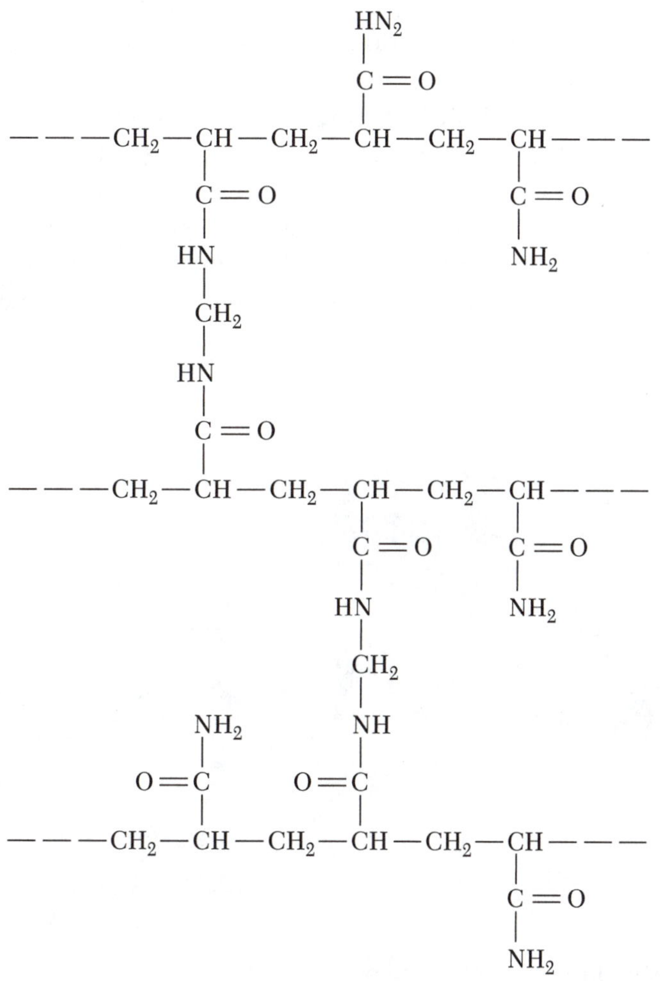

Figure 9.3 The structure of cross-linked polyacrylamide

Amount of Bisacrylamide Cross-Linker

This is not as important as you might think. Most students would assume that to control the pore size of a gel, the cross-linking reagent would be most important. However, it turns out that there is an optimum percentage of cross-linker, between 3 and 5% of the total, that gives the best gels.

TEMED

TEMED is a catalyst that stimulates the formation of free radicals during the reaction that links the acrylamide molecules together. The amount of TEMED controls the speed at which the gel will harden.

Ammonium Persulfate

Ammonium persulfate is the initiator of the reaction. It creates free radicals and begins a chain reaction that links all of the acrylamide molecules together. Like TEMED, the amount of ammonium persulfate controls the speed. We usually use small volumes of highly concentrated ammonium persulfate. You must weigh it out and pipet it carefully. A small error can be the difference between your gel not solidifying during this geological epoch and solidifying in your flask before you get to pour it between the plates.

Many times a discontinuous gel is made in which the bulk of the gel is high percentage at pH 8.5, but a couple of centimeters of gel on top is low percentage (3%) with a pH of 6.5. This upper gel is called the **stacking gel,** because it tends to compress all of the proteins into a thin band. The lower gel is called the **running, resolving,** or **separating** gel because in it the proteins separate from each other, with the small proteins running fastest. The stacking gel works in two ways. First, because it is a low-concentration gel, proteins move quickly in it. When they encounter the higher-density separating gel, they slow down. Therefore, the ones that go into the separating gel first are going slower than the ones still

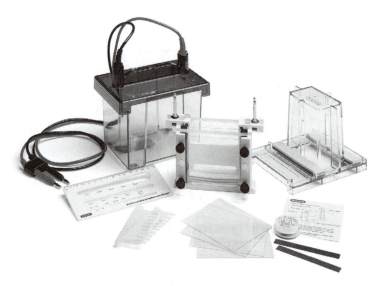

Figure 9.4   Bio Rad Mini Protean II electrophoresis system
(Courtesy of Bio Rad)

in the stacking gel. This causes an accordion effect, and the protein sample becomes much thinner as it enters the separating gel. Second, the pH difference plays upon the charge nature of the compounds being used. Glycine, which is in the buffers, has only a partial negative charge at pH 6.5. This causes zones to be set up in the lane, in which the proteins are sandwiched between chloride ions with lots of negative charge and glycine ions with less charge. This causes voltage differences within the zones that tend to push all of the proteins together. Once the proteins and the glycine enter the separating gel, the pH increase to 8.6 puts more negative charge on the glycine, relieving this effect. The proteins will then move at different rates based on their size.

Many different apparatuses for running polyacrylamide gels are available. A common one is the Bio-Rad Mini Protean® II shown in Figure 9.4.

9.4 SDS-PAGE

As we know, electrophoresis separates proteins based on size, shape, and charge. Native gels are often difficult to interpret because of these three variables. When we wish to use electrophoresis to determine the molecular weight of a protein, we usually do the electrophoresis in the presence of the detergent sodium dodecyl sulfate (SDS), which has the following structure:

$$CH_3-(CH_2)_{11}-SO_4^-$$

SDS binds to proteins in a constant ratio of 1.4 g SDS per gram of protein and covers the protein with negative charges. The SDS and some β-mercaptoethanol included in the sample buffer also denature the proteins and break up any polymers into their subunits. The β-mercaptoethanol reduces any disulfide bridges present. The effect is that all proteins attain the same shape (random coil) and have the same charge/mass ratio. The only variable left is the mass. Proteins, therefore, separate on the gels solely based on molecular weight.

SDS-PAGE is often used to determine if a protein is pure and if there are subunits. If we ran LDH (see Section 9.6) on an SDS gel, the LDH would be broken into its subunits. Instead of having a native tetramer migrating down the gel, we would have the individual M and H monomers migrating. For example, you might run a protein on gel filtration and see a molecular weight of 100,000. If you ran the same protein on SDS-PAGE you might see one band with a molecular weight of 50,000. This would tell you that your protein was made up of two equal-sized subunits. On the SDS-PAGE, you might have seen two bands, with one at 75,000 and the other at 25,000. These add up to 100,000, so you would conclude that you have two unequal-sized subunits in the native molecule.

Standard Curves for Molecular Weight

As stated earlier, one of the main uses of SDS-PAGE is the determination of molecular weight (MW). When the gels are stained and destained, blue bands show up at locations on the gel based on MW. If the log of the MW is graphed vs. the R_m of the protein band (distance the band traveled/distance the tracking dye traveled), a straight line is seen for most proteins. Usually several protein standards are run along with an unknown. The MW of the unknown is calculated by interpolation from the standard curve. Remember to use two-cycle

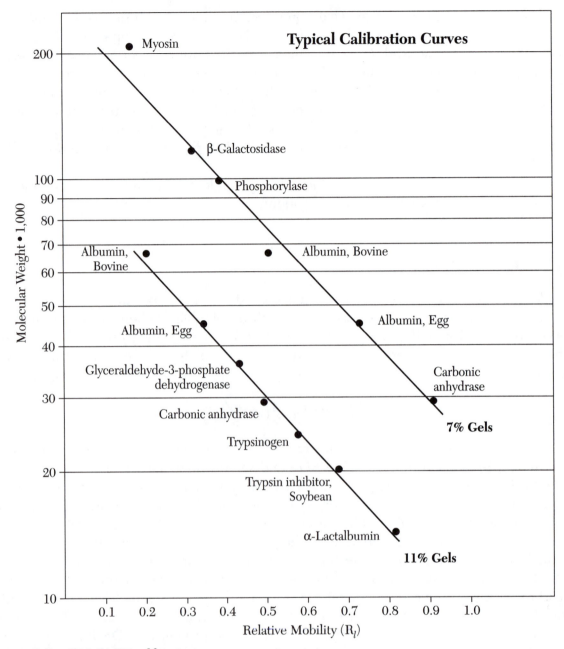

Figure 9.5   SDS-PAGE calibration curves

log paper if your standards have molecular weights that span more than one order of magnitude. Most computer graphing programs can also plot log MW for you. Figure 9.5 shows a typical calibration curve for proteins of known molecular weight with SDS-PAGE.

Tip 9.3

Remember that SDS-PAGE separates proteins based on their size (molecular weight). It also breaks up proteins into subunits. The number you calculate will be the molecular weight of subunits, if any.

9.5 Staining Gels

Many different dyes are used during electrophoresis, and it is important (and often confusing) to keep straight what they all are for. The first type is bromophenol blue, which is included in the sample buffer (also called the tracking dye or loading dye). This acts as a marker so you can see how the separation is proceeding. During the run, it is the only band that you will see. Bromophenol blue is negatively charged and small, so it migrates more quickly than the proteins you are trying to separate. By using bromophenol blue, you can be confident that if the dye has not run off the gel, then neither have your proteins.

The second is Coomassie Blue, which is similar to the dye used in the Bradford protein assay. It is used after an electrophoresis is over and stains all proteins blue. This is used to show you where the proteins are on the gel. Figure 9.6 shows a typical result of a Coomassie gel. Other stains, such as a copper stain or silver stain, are available. They often give more sensitivity and can therefore be used with smaller amounts of protein on the gel.

However, another way to stain a gel is with a chemical specific for a particular protein or enzyme. This is called an activity stain. In Experiment 9, we will use an activity stain that is specific for lactate dehydrogenase. It contains NAD^+, lactate, phenazine methosulfate (PMS_{ox}), and nitroblue tetrazolium (NBT_{ox}), which undergo the following reactions:

$$NAD^+ + Lactate \longrightarrow NADH + Pyruvate$$
$$NADH + PMS_{ox} \longrightarrow NAD^+ + PMS_{red}$$
$$PMS_{red} + NBT_{ox} \longrightarrow PMS_{ox} + NBT_{red}$$

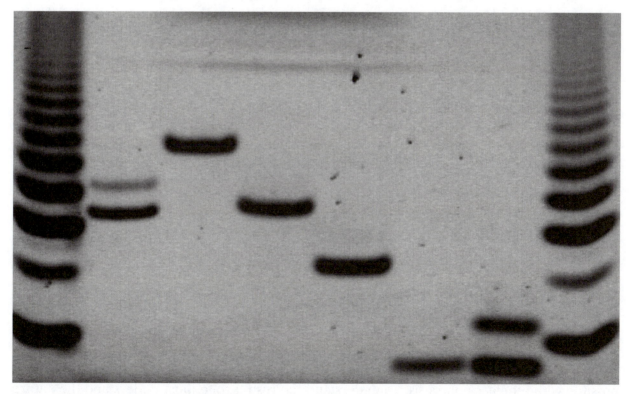

Figure 9.6 Coomassie Blue–stained gel
(Courtesy of Bio Rad)

SDS-PAGE is used to separate proteins solely by their molecular weight. The variables of shape and charge are eliminated by the SDS and by β-mercaptoethanol. All proteins will be negatively charged and run in the correct direction. If the native molecule is made up of smaller subunits, it is the subunits that you will see on the gel. The molecular weight you calculate will be the subunit molecular weight. To determine if a molecule has subunits, we can combine SDS-PAGE with gel filtration chromatography (Chapter 7). The latter would give a larger molecular weight, telling us we had subunits and how many there are.

The first reaction only occurs if lactate dehydrogenase is present. Reduced NBT forms a purple precipitate, so wherever there is LDH in the gel, there will be a purple band.

9.6 Lactate Dehydrogenase

Lactate dehydrogenase (LDH), is a glycolytic enzyme that is present in all tissues. It catalyzes the final step of anaerobic glycolysis, regenerating NAD^+:

$$\text{Pyruvic Acid} + \text{NADH} \rightleftharpoons NAD^+ + \text{L-Lactic acid}$$

LDH is an allosteric enzyme with four subunits and has a total molecular weight of 150,000. There are two types of subunits, H and M, which vary slightly in their amino acid compositions. The H subunit is more prevalent in heart tissue, the M in muscle tissue. These subunits are combined in every possible combination to give four subunits per native LDH molecule (i.e., H_4, H_3M, H_2M_2, HM_3, and M_4). The two types of subunits have different charges, so the five isozymes formed by the different combinations will have different migrations on native gels.

The difference in charge between the various isozymes of LDH is the basis of a sensitive clinical test for myocardial infarctions. The heart muscle contains predominantly the H_4 isozyme, with lesser amounts of H_3M. In blood serum under normal circumstances, the predominant isozymes are H_3M, H_2M_2, and HM_3. After a heart attack, damaged heart cells spill their contents into the blood serum, and the blood serum levels of H_4 and H_3M rise dramatically.

Experiments 9 through 9c comprise the analysis of LDH. In Experiments 9 and 9a, native gel electrophoresis will be used to analyze isozymes of LDH. In Experiments 9b and 9c, SDS-PAGE will be used to determine the purity and subunit molecular weight of LDH.

9.7 Why Is This Important?

This chapter begins our study of electrophoresis, which is perhaps the most-used technique outside of spectrophotometry. Modern biochemistry and molecular biology would be impossible without electrophoresis. To understand the results, one must understand the nature of the separation. Was it based on size, shape, or charge, or a combination of all three? When you see bands on a gel, you have to know what they mean. A native gel does not denature the sample. In Experiments 9 and 9a, the LDH molecules travel in their native conformations and retain all of their activity. This type of electrophoresis is common for purification of proteins and nucleic acids when native conformation is required.

The three biggest fields of study today are protein purification, immunology, and molecular biology. All three use polyacrylamide gels to separate proteins or DNA. A sequencing gel is a large polyacrylamide gel with bands that differ by only one base. The principles for the formation and use of the gel are the same as those in Experiments 9b and 9c. We will use this technique again in Experiment 10 as the first part of the Western blot.

In addition, if you are going into the sciences, agarose gel electrophoresis and SDS-PAGE are two techniques you definitely want on your resume. They will make you much more marketable once your undergraduate career is over.

9.8   Expanding the Topic

Variables and Controls

Science involves setting up experiments that answer a particular question. In many ways, it is a black box. You mix some things together, make some measurements, and get some numbers. What these numbers mean depends on how good your experiment was, whether you ran proper controls, and how sure you are that the numbers you generated reflect what you think they do. To get the most out of your experiments, you must do more than just mix stuff together. Ask yourself, "What's the variable?" and "What controls did we run?"

In Experiments 9 and 9a, we will isolate one variable—charge. Agarose is not the best support for a protein of 150,000 daltons, which is why the bands will be a little diffused. However, we can get all of the information we need because the isozymes separate so nicely from one another. How do we know that we will only separate by charge and that if two bands move different distances they have different charges? We know this because we know that the LDH isozymes all have the same size and shape, which are the other variables with proteins. Without knowing this, we would know nothing from the experiment because we would have three variables. How do we know what isozymes are present in crude bovine heart LDH? We know because we run controls with LDH 1, 2, and 5. If we just run one LDH sample and see three bands, we would not know if they are LDH 1, 2, and 3 or 2, 3, and 4, and so on.

What's This 5× Stuff?

When a buffer or other reagent is given the abbreviation "something ×," this means that the reagent is that many times more concentrated than it should be in its final form. If we make up a 5× reservoir buffer, then it should be diluted 5 times before being put into the electrophoresis chamber. If we have a 2× tracking dye (sample buffer), then we mix 1 part sample buffer to 1 part sample before we run it on the gel. This way, the final concentration in the sample well is correct.

Native Gel Separation of LDH Isozymes (short version)

In this experiment, you will use agarose gel electrophoresis to separate several isozymes of LDH and analyze the charge nature of the H and M subunits.

Prelab Questions

1. How much agarose must you weigh out to make the gel?

2. Describe the order of addition of parts and solutions to the casting tray and the order of removal of the parts from the casting trays.

3. What device will you use to load the samples? Give a possible reason for choosing this over the other options.

Objectives

After successfully completing this experiment, the student will be able to

1. Pour and load an agarose gel for electrophoresis.

2. Identify LDH isozymes based on their electrophoretic migrations.

3. Predict migration patterns for unknown proteins given pIs and molecular weights.

Experimental Procedures

Materials

Fisher electrophoresis chambers and power supplies

Gel trays and combs

Agarose

0.02 M TRIS, 0.02 M glycine, 0.002 M EDTA, pH 8.6 (reservoir buffer)

5× Sample buffer with glycerol and bromophenol blue

LDH activity stain, 0.1 M TRIS, 1% lactate, 0.05% NAD$^+$, 0.0005% PMS (phenazine methosulfate), and 0.005% NBT (nitroblue tetrazolium), pH 9.2

LDH 1, LDH 5, and LDH 2 standards

Crude bovine heart LDH

Crude rabbit muscle LDH

LDH isotrol (LDH 1–5 mixture)

Methods

These procedures are written for Fisher Mini-Submarine gels, and may need to be revised if you have different equipment.

Preparation of Agarose

1. Prepare a mixture of 0.8% w/v agarose in reservoir buffer (TRIS-glycine, pH 8.6). You will need to make 40 mL of the solution in a flask.

2. Heat the flask in a microwave until the agarose has melted. Swirl to mix. This would take 30–40 s on maximum power if only one flask is present.

3. Allow the solution to cool for 3 min before pouring into the casting trays. It should be clear, with no significant agarose clumps.

Pouring the Agarose

1. Insert the rubber stoppers into the casting tray with the thick sides up.

2. Pour the 40 mL of agarose into the casting tray.

3. Insert the comb into the gel. Slide the block holding the comb up against the rubber stopper. Make sure the orientation is correct. The comb itself should not rest against the rubber stopper but should be about a centimeter from it.

4. Let the gel cool for 15 min. Ever so carefully, remove the rubber stoppers before removing the comb. You may have to use a spatula to break the surface tension between the gel and the rubber stopper.

Sample Preparation and Application

1. For each LDH sample to be loaded, take 10 μL of sample and add 2.5 μL of 5× sample buffer. Vortex to mix the contents of the tube. Spin the tubes in a microfuge for a couple of seconds to bring the entire sample back down.

2. Transfer the casting trays to the electrophoresis unit with the sample wells nearest the black (−) electrode.

3. Slowly fill the chamber with reservoir buffer until the buffer is about 1 cm over the top of the gel.

4. Using a Pipetman, apply the samples to the wells. Load the entire sample if it will fit without overflowing, but do not let the solution overflow into the other wells.

5. Put on the lid and connect the leads to the power supply.

6. Set the voltage to 120 V and electrophorese until the marker dye is about 2 cm from the end.

Staining

1. Obtain a small plastic box with a lid. Carefully slide the gel out of the casting tray and into the box.

2. Cover the gel with about 25 mL of LDH activity stain.

3. Incubate the gel for 15–30 min in a 37°C oven or water bath until the bands develop. Remove the gel from the stain when the incubation is done, and place it in water. Sketch or photograph as directed.

Analysis of Results

Experiment 9: Native Gel Electrophoresis (short version)

Data

Describe the results of your separation by including a sketch or photo of the gel. Be sure to correctly label the lanes with which sample was applied. Sketch it so that the wells are at the top.

Analysis of Results

1. Why do some of the lanes have more than one band in them? What does each band represent?

2. Describe the nature of the LDH molecules that are separated on this system. How do we know that there is only one variable at work here?

Questions

1. Describe the effects the following would have on the way your gel would run. Use this equation to justify your answer:

$$v = \frac{qE}{f}$$

    a. Increasing the agarose concentration to 1.5%.

b. Increasing the pH of the TRIS-glycine buffer to 9.5.

c. Decreasing the voltage at which the gel was run.

2. You need to load 20 μg of protein into one of the wells of a gel. This needs to be in 1×
    buffer and in a total volume of 30 μL. You are given a 10 μg/μL solution of protein, a
    10× buffer, and water. How much of each should you mix together in order to load the
    gel correctly?

Experiment 9a

Native Gel Separation of LDH Isozymes (comprehensive)

In this experiment, you will use agarose gel electrophoresis to separate the isozymes of LDH, determine which ones are in your purified samples, and analyze the charge nature of the H and M subunits.

Objectives

After successfully completing this experiment, the student will be able to

1. Pour and load an agarose gel for electrophoresis.

2. Identify LDH isozymes based on their electrophoretic migrations.

3. Predict migration patterns for the purified LDH isozymes.

Experimental Procedures

Materials

Fisher Mini-Submarine gel apparatuses

Power supplies

Agarose

0.02 M TRIS, 0.02 M glycine, 0.002 M EDTA, pH 8.8.

5× Sample buffer (0.2 M TRIS, 0.2 M glycine, 0.01 M EDTA, 0.02% bromophenol blue, 25% glycerol, pH 8.8)

LDH activity stain, 0.1 M TRIS, 1% lactate, 0.05% NAD^+, 0.0005% PMS, and 0.005% NBT, pH 9.2.

LDH 1, LDH 5, and LDH 2 standards

Crude bovine heart LDH

LDH isotrol (contains all five isozymes)

Your LDH samples

Procedures

These procedures were written for the Fisher Mini-Submarine gels. They may need to be slightly modified for other equipment.

Preparing Agarose

1. Prepare a mixture of 0.8% w/v agarose in TRIS-glycine pH 8.6 buffer. You will need to make 40 mL of the solution.

2. Heat the flask in a microwave until the agarose has melted. Swirl the flask to mix the contents. This would require 30–40 s on high power for a single flask in the microwave.

3. Allow the solution to cool for 1 min before pouring it into the casting trays. It should be clear with no significant agarose clumps.

Pouring the Agarose

1. Insert the rubber stoppers into the casting tray with the thick sides up.

2. Pour the 40 mL of agarose into the casting tray.

3. Insert the comb into the gel. Slide the block holding the comb up against the rubber stopper. Make sure the orientation is correct. The comb itself should not rest against the rubber stopper but should be about a centimeter from it.

4. Let the gel cool for 15 min. Ever so carefully, remove the rubber stoppers before removing the comb. You may have to use a spatula to break the surface tension between the agarose and the rubber.

Sample Preparation and Application

1. For each LDH sample to be loaded, take 10 μL of sample and add 2.5 μL of 5× sample buffer. Use the vortexes to mix the contents of the tube. Spin the tubes in the microfuges for a couple of seconds to bring the entire sample back down. The samples should be loaded as follows:

   Lane 1: LDH 1 isozyme

   Lane 2: LDH 2 isozyme

   Lane 3: LDH 5 isozyme

   Lane 4: Crude bovine heart LDH

   Lane 5: LDH isotrol (LDH 1–5)

   Lane 6: Your best sample

   Lane 7: Another of your samples

   Lane 8: Another of your samples

2. Transfer the casting trays to the electrophoresis unit with the sample wells nearest the black (−) electrode.

3. Slowly fill the chamber with electrophoresis buffer until the buffer is about 1 cm over the gel.

4. Using a Pipetman, apply the samples to the wells. Load the entire sample if it will fit without overflowing, but do not let the solution overflow into the other wells.

5. Put on the lid and connect the leads to the power supply. Electrophorese at 100 V until the marker bromophenol blue dye is close to running off of the end. At 125 volts, the separation would take about an hour and a quarter. However, the gel will get very hot and the bands will not give as clean a separation. At 80 volts, the separation will take 2–3 h, but the results will be better.

Staining

1. Obtain a small plastic box with a lid. Carefully slide the gel out of the casting tray and into the box.

2. Cover the gel with about 25 mL of LDH activity stain.

3. Incubate the gel for 5–15 minutes in a 37°C water bath until the bands develop.

4. Photograph or sketch the gel.

Analysis of Results

Experiment 9a: Native Gel Electrophoresis (comprehensive)

Data

Describe the results of your separation by including a sketch or photo of the gel. Be sure to correctly label the lanes with which sample was applied there. Sketch it so that the wells are at the top.

Analysis of Results

1. Why do some of the lanes have more than one band in them? What does each band represent?

2. Describe the nature of the LDH molecules that are separated on this system. How do we know that there is only one variable at work here?

3. Which of your samples was the most active? Did you get the isozyme pattern you expected? Why or why not?

Questions

1. Describe the effects the following would have on the way your gel would run. Use this equation to justify your answer:

$$v = \frac{qE}{f}$$

    a. Increasing the agarose concentration to 1.5%.

    b. Increasing the pH of the TRIS-glycine buffer to 9.5.

    c. Decreasing the voltage at which the gel was run.

2. You need to load 20 μg of protein into one of the wells of a gel. This needs to be in 1× buffer and in a total volume of 30 μL. You are given a 10 μg/μL solution of protein, a 10X buffer, and water. How much of each should you mix together in order to load the gel correctly?

Experiment 9b

SDS–Polyacrylamide Gel Electrophoresis (short version)

In this experiment, SDS-PAGE will be used to separate proteins by molecular weight.

Prelab Questions

1. What molecular weight should you find for the LDH band?
2. What band of your standards will it be closest to?
3. Describe the procedure to use after the stacking gel has hardened and before you load your samples.

Objectives

After successfully completing this experiment, the student will be able to

1. Pour polyacrylamide gels and assemble the gel chamber.
2. Prepare and load protein samples in the sample wells.
3. Stain and destain the gels with Coomassie Blue.
4. Determine the molecular weights of unknown proteins.

Experimental Procedures

Materials

Separating gel and stacking gel solutions

Reservoir buffer

BSA, 1 mg/mL (MW 66K)

Ovalbumin, 1 mg/mL (MW 45K)

Glyceraldehyde-3-phosphate dehydrogenase, 1 mg/mL (MW 36K)

Carbonic anhydrase, 1 mg/mL (MW 29K)

Trypsinogen, 1 mg/mL (MW 24K)

Dalton Mark VII Standard Proteins mixture (1 mg/mL each BSA, ovalbumin, glyceraldehyde-3-phosphate dehydrogenase, carbonic anhydrase, trypsinogen, trypsin inhibitor, MW 20K, and lactalbumin, MW 14K)

Unknown proteins

Methods

Gel Preparation

These procedures are for the Bio Rad Mini Protean II. Your procedures may need to be revised to accommodate a different setup.

> **Tip 9.4**
> ⟲ *CAUTION!* Acrylamide is a neurotoxin. It is dangerous when liquid, so wear gloves until both of your gels are between the plates and solid.

1. Set up the gel plates per Bio Rad instructions. Do this part very carefully. If you do not get the plates lined up properly or do not use the plate leveling part of the casting tray, the gel will leak when you pour it.

2. After you have leveled the plates and set the plate assembly in the casting tray, put the comb in and mark the plate on the outside 2 mm below the bottom of the comb. The bridge of the comb should come to rest on top of the gray plate spacers.

3. Prepare 1 mL of 10% w/v ammonium persulfate in water and store on ice. Note that 1 mL would be enough for the entire class.

4. Use a Pipetman to add 17 μL of the ammonium persulfate (AP) to 5 mL of separating gel. Swirl the flask gently to mix the contents and quickly inject the solution between the plates up to the level of the mark that you made. This would be a bad time to discover that your pipet of choice does not fit into the vessel containing the solidifying acrylamide. A liquid transfer syringe or 5-mL pipet with a pipet pump would work well here.

5. If no leaks occur, use a Pasteur pipet to layer a few millimeters of butanol on top of the gel.

6. When the gel has hardened, remove the butanol and wash with reservoir buffer.

7. Add 13 μL of AP to 2.5 mL of the stacking gel and swirl gently.

8. Inject the stacking gel to the top of the shorter glass plate and insert the comb to the point you marked before. Try to avoid having air bubbles stuck to the comb.

9. When the stacking gel has hardened (15 min), remove the comb and wash the wells with reservoir buffer. Place the glass plate assembly into the central clamp assembly. Fill the upper (central) reservoir as full as possible and check for leaks. Fill the tank with

reservoir buffer up to the level of the lower electrode. The gel is now ready for sample application.

Sample Preparation and Application

1. Boil each protein standard or unknown listed for 5 min, then place it on ice. Spin the tubes so the entire sample is at the bottom. These samples have already had the sample buffer/tracking dye added. Boiling denatures the proteins and allows the SDS and β-mercaptoethanol to react.

2. Using the 25-μL Hamilton syringes or Pipetmen, load 10 μL of the samples into the wells. Load slowly, allowing the solution to layer on the bottom of the well. *This is your last chance to avoid making the common error of loading samples before filling the reservoir with buffer.*

3. Load the samples in this order:

   Lane 1: Sample buffer only

   Lane 2: Dalton VII mixed markers (contains all of the markers)

   Lane 3: BSA, MW 66,000

   Lane 4: Ovalbumin, MW 45,000

   Lane 5: Glyceraldehyde-3-phosphate dehydrogenase, MW 36,000

   Lane 6: Carbonic anhydrase, MW 29,000

   Lane 7: Trypsinogen, MW 24,000

   Lane 8: Unknown protein

   Lane 9: LDH

   Lane 10: Dalton VII mixed markers

   If you have problems with one of the lanes, you can use the outside lane to repeat that sample.

   When you are changing samples, rinse the Hamilton syringes three times with 0.1% SDS or use a fresh pipet tip.

4. Now you should be ready for electrophoresis, but before you begin, check to make sure you didn't have a slow leak from the upper reservoir chamber to the lower. If the upper chamber now has less buffer in it, you will have to add more buffer to the top. One way to minimize the effects of leaking is to add more buffer to the lower chamber until heights almost equal the height the lower buffer and the upper buffer have.

Electrophoresis

1. Put the top on the unit, connecting the red leads to the red plugs on the power supply and the black to the black.

2. Have a TA or the instructor check the apparatus.

3. Electrophorese at 200 V until the dye front reaches the bottom (30–45 min).

4. Label a plastic container for staining and destaining.

> **Tip 9.6**
>
> Before staining your gel, make sure you can identify which side is left and which is the top. Once the gel has rolled around in Coomassie Blue for awhile, this might not be so easy.

5. Separate the plates and loosen the gel from the plate by squirting water under it with a liquid transfer syringe. Place the gel into the plastic container and overlay with Coomassie Blue protein stain.

6. The gel will be stained overnight and then placed into destain.

7. To view the gel, place the gel on the lid, not on a paper towel.

Analysis of Results

Experiment 9b: SDS-PAGE (short version)

Data

Unknown # _____

1. Make a sketch or take a picture of your gel labeling the protein lanes.

2. Calculate the R_ms for the bands. This is best done using the Dalton VII lane. The single standard lanes are used to verify the identity of a band in the Dalton.

Protein	R_m
BSA	_____
Ovalbumin	_____
Glyceraldehyde-3-phosphate dehydrogenase	_____
Carbonic anhydrase	_____
Trypsinogen	_____
Unknown	_____
LDH	_____

Analysis of Results

1. Plot log MW vs. R_m for the bands, and turn in the graph with this report.

2. Determine the MW of the unknown.

Questions

1. The last gel we did separated proteins based on charge. What is the variable this time? How do we know there is only one variable at work here?

2. You have an enzyme that is composed of three subunits. Two of the subunits are 28 kD and the other is 14 kD. How many bands will you see on an SDS-PAGE?

3. Describe the results you would see if you did an experiment to determine the physical properties of tyrosinase. If you ran gel filtration and SDS-PAGE, what results would you see?

Experiment 9c

SDS-Polyacrylamide Gel Electrophoresis (comprehensive)

In this experiment, SDS-PAGE will be used to separate proteins by molecular weight and to verify the purity of your samples.

Objectives

After successfully completing this experiment, the student will be able to

1. Pour polyacrylamide gels and assemble the gel chamber.

2. Prepare and load protein samples in the sample wells.

3. Stain and destain the gels with Coomassie Blue.

4. Determine the molecular weights of unknown proteins.

5. Verify the purity of your best LDH samples.

Experimental Procedures

Materials

Acrylamide/bisacrylamide*

TEMED†

4× Separating buffer

4× Stacking buffer

Reservoir buffer

2× Sample buffer†

---

*Acrylamide is a neurotoxin. Wear gloves at all times when handling it or any solution containing it.
†TEMED and the sample buffer will be used in the hood.

Dalton Mark VII protein mixture (1 mg/mL each BSA, ovalbumin, glyceraldehyde-3-phosphate dehydrogenase, carbonic anhydrase [MW 29K], trypsinogen [24K], trypsin inhibitor [20K], and lactalbumin [14K]

BSA, 1 mg/mL (MW 66K)

Ovalbumin, 1 mg/mL (MW 45K)

Glyceraldehyde-3-phosphate dehydrogenase, 1 mg/mL (MW 36,000)

Bio Rad Mini Protean II electrophoresis apparatuses

Power supplies

Procedures

These procedures were written for the Bio Rad Mini Protean II electrophoresis unit. You may need to modify them for different equipment.

Gel Preparation

1. Prepare 10 mL of separating gel and 5 mL of stacking gel by mixing ingredients according to the following table. Do not add the ammonium persulfate yet. Store the gel solutions on ice. Note that this is enough for two gels.

Ingredients for Gels

Reagent	Volume (12% separating gel)	Volume (5% stacking gel)
4× Separating buffer	2.5 mL	None
4× Stacking buffer	none	1.25 mL
dd H$_2$O	5.1 mL	3.25 mL
Acrylamide/bisacrylamide	2.4 mL	0.5 mL
TEMED	6.7 μL	7.5 μL
10% Ammonium persulfate	33 μL	25 μL

2. Clean the plates with great obsessiveness. Set up the gel per Bio Rad instructions. The initial setup in the casting tray will determine whether your gel leaks or not.

3. Prepare 1 mL of 10% ammonium persulfate and store on ice. Note that this volume is sufficient for the entire class.

4. Place the comb into the gel plates and insert until the outermost stop comes to rest on top of the gray spacers. With a Sharpie™ pen, mark the glass plate at the bottom of the comb.

5. Divide the separating gel into two 5-mL aliquots and add 17 μL of the ammonium persulfate (AP) to 5 mL of the separating gel. Swirl the mixture gently but thoroughly, and quickly inject the solution between the plates up to the level of the mark that you made.

6. If no leaks occur, use a Pasteur pipet to layer a few mm of butanol onto the gel. Wait until the gel has hardened (15 min).

7. Remove the buffer and butanol that are on top of the separating gel and wash several times with electrode buffer.

8. Divide the stacking gel into two 2.5-mL aliquots and add 13 μL of AP to one of them. Vortex the mixture gently.

9. Inject the stacking gel to the top of the shorter plate and insert the comb. Try to avoid having air bubbles stuck to the comb.

10. When the stacking gel has hardened (15 min), remove the comb. Remove the plate assembly from the casting tray and attach it to the central block. Place the central unit in the tank and fill the upper reservoir with reservoir buffer. Check for leaks. The upper reservoir should be filled to the top of the outer plates. The lower reservoir must be filled to the height of the lower electrode wire, about 1 cm from the bottom of the gel. The gel is now ready for sample application. If buffer leaks from the upper reservoir into the lower, the problem can be diminished by raising the level of the buffer in the lower reservoir.

Sample Preparation and Application

1. Mix 10 μL of each unknown sample with 10 μL of 2× sample buffer in a microcentrifuge tube. The standards have already been mixed with the 2× sample buffer, so just pipet 20 μL into the microcentrifuge tube (40 μL for the Dalton).

2. Boil for 2 min, then place on ice. Spin the tubes so that the entire sample is at the bottom.

3. Using the 25 μL Hamilton syringes or Pipetmen, load 15 μL of the samples into the wells. Load slowly, allowing the solution to layer on the bottom of the well.

4. Load the samples in the following order:

    Lane 1: Dalton mixed markers

    Lane 2: BSA

    Lane 3: Ovalbumin

    Lane 4: Glyceraldehyde-3-phosphate dehydrogenase

    Lane 5: Your 20,000 × g supernatant

    Lane 6: Your 65% ammonium sulfate pellet

    Lane 7: Your pooled Q-Sepharose fraction

    Lane 8: Your pooled Cibacron Blue fraction

    Lane 9: Your pooled gel filtration fraction

    Lane 10: Dalton mixed markers

When changing samples, rinse the Hamilton syringes three times with 0.1% SDS or use a fresh pipet tip.

Electrophoresis

1. Put the top on the unit, connecting the red leads to the red plugs on the power supply and the black to the black.

2. Have a TA or the instructor check the apparatus.

3. Electrophorese at 180 V until the dye front reaches the bottom (45 min). Using a lower voltage will improve the results as it did with the agarose gels.

4. Label (with tape) a plastic container for staining and destaining. Remove the gel from the plates by injecting water with a liquid transfer pipet between the gel and plate. Place the gel in the tray with staining solution. The teaching staff may put the gel in the destain for you or you may have to come back and do it yourself. You need to come in later to observe the results.

Name ______________________________ Section __________

Lab partner(s) ______________________________ Date __________

Analysis of Results

Experiment 9c: SDS-PAGE (comprehensive)

Data

1. Make a sketch or take a picture of your gel labeling the protein lanes.
2. Calculate the R_ms for the bands. This is best done using the Dalton VII lane. The single standard lanes are used to verify the identity of a band in the Dalton.

Protein	R_m
BSA	________
Ovalbumin	________
Glyceraldehyde-3-phosphate dehydrogenase	________
Carbonic anhydrase	________
Trypsinogen	________
LDH	________

Analysis of Results

1. Plot log MW vs. R_m for the bands, and turn in the graph with this report.

2. Determine the MW of the LDH monomer.

3. Is the molecular weight of the LDH monomer what you would expect? Why or why not?

4. Which of your samples was the most pure? Is this what you expected?

Additional Problem Set

1. Explain the purpose of the components of an acrylamide gel.

2. Which components of an acrylamide gel affect the qualities of the final product, and which ones affect the speed of polymerization?

3. If you ran Dalton VII mixed markers on an acrylamide gel and only five bands showed up, what could be some explanations for what happened to the other two?

4. What is the purpose of running both Dalton VII mixed markers and lanes with individual markers on SDS-PAGE?

5. Given the results that you saw with the Dalton markers, is it possible to use a 12% separating gel to run a set of proteins that range in molecular weight from 10,000 to 200,000? Explain.

6. What could cause a protein to migrate on SDS-PAGE in such a way that it appeared to be a protein that was much bigger?

Webconnections

For a list of Web sites related to the material covered in this chapter, go to **Webconnections** at the *Experiments in Biochemistry* site on the Saunders College Publishing Web page. You can access this page at http://www.saunderscollege.com. **Webconnections** are under *Experiments in Biochemistry* in the Biochemistry portion of the Chemistry page.

References and Further Reading

G. Acquaah, *Practical Protein Electrophoresis for Genetic Research,* Dioscorides Press, 1992.

R. C. Allen, *Gel Electrophoresis of Proteins and Nucleic Acids: Selected Techniques,* W. de Gruyter, 1994.

J. N. Anderson, *A Laboratory Course in Molecular Biology,* vol. I, Physiology and Clinical Biology of Proteins, Modern Biology, 1986.

D. M. Bollag, M. D. Rozycki, and S. J. Edelstein, *Protein Methods,* Wiley-Liss, 1996.

R. F. Boyer, *Modern Experimental Biochemistry,* Addison-Wesley, 1993.

R. D. Cahn, N. O. Kaplan, L. Levine, and E. Zwilling, (1962), *Science,* **136.** Nature and Development of Lactic Dehydrogenases.

P. Camilleri, *Capillary Electrophoresis: Theory and Practice,* CRC Press, 1993.

M. Campbell, *Biochemistry,* Saunders College Publishing, 1998.

J. Clausen and B. Ovlisen, (1965), *Biochem. J.* **97.** Lactate Dehydrogenase Isoenzymes of Human Semen.

L. G. Davis, W. M. Kuehl, and J. F. Battey, *Basic Methods in Molecular Biology,* Appleton and Lange, 1994.

R. E. Dickerson and I. Geiss, *The Structure and Action of Proteins,* Harper and Row, 1969.

R. L. Dryer and G. F. Lata, *Experimental Biochemistry,* Oxford University, 1989.

F. Foret, L. Krivankova, and P. Bocek, *Capillary Zone Electrophoresis,* VCH Publishers, 1993.

D. M. Gersten, *Gel Electrophoresis: Proteins—Essential Techniques,* Wiley, 1996.

D. M. Hawcroft, *Electrophoresis,* IRL Press at Oxford University Press, 1997.

P. Jones, *Gel Electrophoresis: Nucleic Acids,* John Wiley and Sons, 1995.

P. B. Kaufman, *Handbook of Molecular and Cellular Methods in Biology and Medicine,* CRC Press, 1995.

G. P. Manchenko, *Handbook of Detection of Enzymes on Electrophoretic Gels,* CRC Press, 1994.

R. Martin, *Gel Electrophoresis,* Oxford Scientific, 1996.

D. Patel, *Gel Electrophoresis,* Wiley, 1994.

T. Rabilloud, (1994), *Cel. Mol. Biol.,* **40.** Silver Staining of Proteins in Polyacrylamide Gels: A General Overview.

J. F. Robyt and B. J. White, *Biochemical Techniques,* Brooks Cole, 1990.

L. E. Schoeff and R. H. Williams, *Principles of Laboratory Instruments,* Mosby-Year Book, 1993.

J. M. Walker, *Basic Protein and Peptide Protocols,* Humana Press, 1994.

K. Weber and M. Osborn, (1969), *J. Biol. Chem.,* **244(16).** The Reliability of Molecular Weight Determinations by SDS-PAGE.

R. Westermeier, *Electrophoresis in Practice: A Guide to Theory and Practice,* (1993), VCH Publishers.

D. H. Whitmore, *Electrophoretic and Isoelectric Focusing Techniques in Fisheries Management,* CRC Press, 1990.

Chapter 10

Western Blots

Topics

Introduction

This chapter discusses the popular technique of Western blotting. A Western blot is a transfer of proteins from an electrophoresis gel (usually polyacrylamide) onto a thin membrane of nitrocellulose or some other absorbent material. Antibodies are then used to locate desired proteins so that their position on the original gel can be determined.

10.1   Western Blot Theory

In the native gel electrophoresis experiment, we separated LDH isozymes on a gel and then identified their locations via an activity stain. That was one way to pick a desired protein out of a mixture, because many of the samples we separated were crude and contained other proteins. What do you do if the desired protein is not an enzyme? There will be no activity stain you can use. The answer is to use antibodies and the process called Western blotting.

In a Western blot, you start by running a regular protein-separating gel, either native or SDS. Once the proteins are separated, you make a blot with the gel and a membrane such as nitrocellulose that binds proteins. The gel/membrane blot is then put into a different type of electrophoresis apparatus that will transfer the proteins out of the gel and onto the membrane. Once this is done, specific antibodies are used to find the protein of interest on the

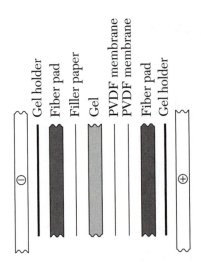

Figure 10.1   Western blot sandwich

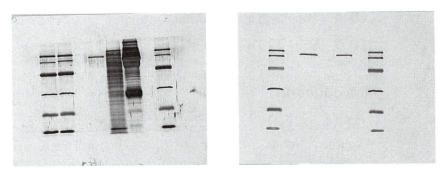

Figure 10.2   Coomassie Blue total protein–stained gel (a) and Western blot (b)
(Courtesy of Bio Rad)

membrane. Figure 10.1 shows the gel/membrane sandwich. Figure 10.2 shows typical results of transferring proteins to a membrane.

Whether you use a native gel or an SDS gel will depend on the antibodies you will use later. Some antibodies react to an epitope that is only present in the native conformation. Other antibodies will react to a protein even if it is completely denatured by SDS. This happens when the antibody reacts to a small linear amino acid sequence.

10.2   Antibodies

The key to the Western blot technique lies in the specificity of the antibodies chosen. Once the proteins are transferred onto the nitrocellulose, they are reacted with two different types of antibodies. The first reaction is with what we call a primary antibody. This is an antibody against the protein we are looking for. It may be a monoclonal antibody, which is very pure and expensive. These are derived from a single B cell, which produces only one type of specific antibody.

Polyclonal antibodies are a mixture collected from serum and are not as specific or expensive. This is shown in Figure 10.3 as number 2. Wherever the primary antibody finds

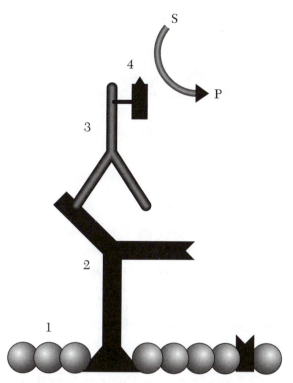

Figure 10.3   Antibody reactions in Western blots

the protein of interest on the membrane, it will bind. Unfortunately, we would not be able to see those results, so we have to do another step. The next step is to react the blots with a secondary antibody. The secondary antibody is an antibody against the general class of primary antibodies. For instance, if we use a mouse primary antibody against human LDH, then our secondary antibody could be a goat antibody against mouse antibodies. The secondary antibody will bind to the blot where there is bound primary antibody. This is indicated in Figure 10.3 as number 3.

10.3   Color Development

The protein/primary antibody/secondary antibody complex would not, in itself, be visible either. However, most secondary antibodies are conjugated to an enzyme or other available marker. Common enzymes are alkaline phosphatase or horseradish peroxidase (HRP). This is shown as number 4 in Figure 10.3. These enzymes catalyze reactions that leave visible products. We eventually incubate the membrane in the substrates for the conjugated enzymes and get a visible band. Horseradish peroxidase was the first label used with secondary antibodies. Horseradish peroxidase–labeled antibodies catalyze the reaction

$$\textbf{4-Chloro-1-naphthol}_{\text{red}} + \textbf{H}_2\textbf{O}_2 \longrightarrow \textbf{4-Chloro-1-naphthol}_{\text{ox}} + \textbf{H}_2\textbf{O} + \tfrac{1}{2}\textbf{O}_2$$

where the oxidized form of naphthol forms a purple precipitate. This secondary antibody gives very clean results with little background staining, but the bands are very light and often difficult to see.

Alkaline phosphatase is another common enzyme tag used with secondary antibodies. It catalyzes the reaction

5-Bromo-4-chloro-3-indolyl phosphate + Nitroblue tetrazolium$_{ox}$ $\longrightarrow$
5-Bromo-4-chloro-3-indole + P_i + Nitroblue tetrazolium$_{red}$

where the reduced form of NBT forms a purple precipitate, as it did with the LDH-activity stain we used in Experiment 9a.

10.4 Blocking and Washing

Interspersed between the antibody reactions are some other steps. If we transferred the proteins onto the nitrocellulose and immediately reacted them with the primary antibodies, they would bind all over the nitrocellulose because this support has a high affinity for all proteins. Blocking is a procedure we use after we do the transfer. It entails suspending the membrane in a solution of nonspecific proteins that will then bind up all of the sites on the nitrocellulose that have not already been bound by the transferred proteins. Common blocking reagents are milk, BSA, or gelatin.

After each incubation with an antibody, there is an extensive washing procedure to get rid of all antibodies that are not bound tightly to the transferred proteins. Without such washing, extensive background color will be seen when the color development solution is used, and it will be difficult to see the bands.

10.5 Why Is This Important?

The importance of molecular biology and immunology are made more apparent every day as scientists fight the AIDS virus. Techniques that allow the study of DNA, and the proteins that are produced from it, grow in importance. There are several blotting techniques used routinely. Southern blots are used to transfer DNA, Northern blots for RNA, and Western blots for proteins. All three of these techniques are similar, and all are based on separating molecules on a gel and then transferring them onto nitrocellulose. Specific molecules can then be identified by reacting the blots with a probe. For Northern or Southern blots, this probe is a sequence of nucleotides. For Western blots the probe is an antibody specific for the protein of interest. Once the location of the protein is established, it can be excised from the gel and studied further.

ESSENTIAL INFORMATION

Western blot is a technique for identifying and quantifying a particular protein out of a mixture. Typically, an acrylamide gel is used to separate proteins according to molecular weight or charge via SDS-PAGE or native gel electrophoresis. Then the gel is put into a sandwich and a second electrophoresis drives the proteins out of the gel and onto a membrane such as nitrocellulose. Once the proteins are isolated and concentrated on the membrane, very specific antibodies are used to probe for the presence of the protein of interest. A very specific primary antibody binds to the protein. A secondary antibody then binds to the primary antibody. This secondary antibody carries an enzyme or fluorescent marker that allows the protein–antibody complex to be seen.

Western Blots of Serum Proteins

In this experiment, you will use native gel electrophoresis on polyacrylamide to separate serum proteins from several species. The separated proteins will be transferred to nitrocellulose via the Western blot procedure. The nitrocellulose will then be incubated with antibodies to locate the position of human serum albumin.

Prelab Questions

1. Briefly diagram the order of events for this experiment.

2. Indicate how you will mark the gels and blots so that you know which bands pertain to which sera.

Objectives

Upon successful completion of this experiment, the student will be able to

1. Set up a Western blot apparatus and gel sandwich.

2. React blots with primary and secondary antibodies.

3. Perform color development reactions.

4. Identify desired proteins by comparing blots to acrylamide gels.

Experimental Procedures

Materials

Acrylamide gels

2× Sample buffer

Serum samples

Power supplies

Transfer buffer

Nitrocellulose

Coomassie Blue R protein stain

TBST (0.05% Tween 20 in TBS)

Milk solution (10% milk powder in TBST)

Mouse anti-HSA

Goat anti-mouse IgG

37°C Water baths

Color developing solution

Methods

Lab Period one

1. Acquire an electrophoresis apparatus with prepoured gel.

2. Mix each sample with an equal volume of 2× sample buffer (SDS-free). Load each lane with the serum samples, about 10 μL per lane. Load the 10 lanes of the gel in this order:

   Lanes 1, 6: BSA

   Lanes 2, 7: Human serum

   Lanes 3, 8: Pig serum

   Lanes 4, 9: Cow serum

   Lanes 5, 10: Horse serum

3. Electrophorese at 200 V as before until the dye front is close to the bottom of the gel.

4. Cut the gel in half lengthwise so that you have two gels with five lanes each. Make a notch in one of the corners, so that you know the proper orientation.

5. Place one half into a tray with Coomassie Blue R protein stain.

6. Wet all materials for the gel sandwich with transfer buffer. Wear gloves during all handling of the gel, blotting papers, and nitrocellulose.

7. Build the gel sandwich with blotting paper and nitrocellulose with two pieces of filter paper, the gel, one piece of nitrocellulose, and two more pieces of filter paper. All should be the same size as the gel. Put a notch in the nitrocellulose so that you know its orientation. Place the sandwich in the transfer apparatus, making sure the nitrocellulose is toward the + electrode.

8. Prepare a plastic tray with 30 mL of blocking solution. Label the tray with your name.

9. At some point the next day, the Coomassie Blue–stained gel must be put into destain solution (10% acetic acid). This will later be used to see where all of the proteins are.

Lab Period 2

1. Wash the nitrocellulose membrane a few times with distilled water and then with TBST. Pour off the last washing of TBST.

2. Place the nitrocellulose membrane into a tray with 20 mL of primary antibody. Incubate the tray for 30 min on a shaker.

3. Wash the membrane three times for 10 min each in 20 mL of TBST to remove unbound antibody.

4. Replace the TBST with 20 mL of secondary antibody.

5. Incubate for 30 min as before.

6. Pour off the secondary Ab-HRP conjugate and wash three times for 10 min each with 20 mL of TBST.

7. Blot the membrane damp dry on filter paper and transfer it to 20 mL of the color development solution. Reactive areas will turn purple in about 15 min, but reaction will continue for up to 4 h.

8. When the blot has developed to a desired intensity, stop the reaction by rinsing the membrane with distilled water. The membrane can be air dried for storage on filter paper. Protect from light. The band will fade over time.

9. Compare the nitrocellulose to the Coomassie Blue–stained gel.

Analysis of Results
Experiment 10: Western Blots

Data

Draw a picture of your gel and blot showing the location of the bands and the identification of the lanes. Use a separate sheet of paper or turn in a photograph if you have one.

Calculations and Questions

1. Which of the bands on the native gel represents serum albumin?

2. From this experiment, what can you say about the similarities and differences between serum albumin and the different species tested?

3. Suppose that upon developing your blot, you see more than one band. What would be some possible reasons for this?

4. Give the mechanism/purpose of the following steps in the blotting procedure. Why did you do it and/or how does it work?

a. Tagging the secondary antibody with horseradish peroxidase.

b. Soaking in milk solution.

c. Running a native acrylamide gel.

5. How could you test the efficiency of transfer of the proteins from the gel to the nitro-cellulose membrane?

Experiment 10a

Western Blot of LDH

In this experiment, you will separate your LDH samples on a native polyacrylamide gel, transfer them to nitrocellulose via the Western blot procedure, and then probe for LDH using an anti-human LDH-H subunit antibody.

Objectives

Upon successful completion of this experiment, the student will be able to

1. Set up a Western blot apparatus and gel sandwich.

2. React blots with primary and secondary antibodies.

3. Perform color development reactions.

4. Identify desired proteins by comparing blots to acrylamide gels.

Experimental Procedures

Separation of LDH isozymes (day 1)

Materials

Acrylamide/bisacrylamide*

TEMED†

4× Separating buffer (SDS-free)

4× Stacking buffer (SDS-free)

TRIS-glycine reservoir buffer (SDS-free)

2× sample buffer (SDS-free)†

Bio Rad Mini Protean II electrophoresis apparatuses

Power supplies

Human LDH sample

Bovine LDH sample

---

*Acrylamide is a neurotoxin. Wear gloves at all times when handling it or any solution containing it.
†TEMED and the sample buffer will be used in the hood.

Human serum

Your LDH samples

TRIS-glycine, MeOH transfer buffer

TRIS buffered saline (TBS)

TRIS buffered saline + Tween 20 (TBST)

Genie transblotters

Procedures

1. Prepare a 12% acrylamide gel (as in Experiment 9c but SDS-free).

2. Mix each sample with an equal volume of 2× sample buffer (SDS-free). Load each lane with the 15 μL samples. Load the ten lanes of the gel so that the first five lanes contain one lane each of the five different samples and the second five lanes also contain one lane each of the samples.

3. Electrophorese as in Experiment 9b until the dye front runs off the gel.

4. Cut the gel in half lengthwise so that you have two gels with five lanes each. Make a notch in one of the corners, so that you know the proper orientation.

5. Place one half into a tray with Coomassie Blue R protein stain.

6. Wet all materials for the gel sandwich with transfer buffer (TRIS-glycine reservoir buffer with 20% methanol). Wear gloves during all handling of the gel, blotting papers, and nitrocellulose.

7. Build the gel sandwich with blotting paper and nitrocellulose so that there are two pieces of filter paper, the gel, one sheet of nitrocellulose, and two more pieces of filter paper. Put a corresponding notch in the nitrocellulose so that you know its orientation. Place the sandwich in the electroblotting apparatus so that the nitrocellulose is toward the + electrode. Place a labeled container of blocking solution (30 mL of 10% milk powder in TBS + Tween 20, known from now on as TBST) near the apparatus.

8. Transfer the proteins onto the nitrocellulose at 25 V for 1–2 h, and put your blot into the tray with blocking solution (this step may be done for you). The gel that was used for the transfer should also be stained with Coomassie Blue.

9. Incubate for 1 h to one week.

10. At some point, the Coomassie Blue–stained gels must be put into destain solution (10% acetic acid).

Reaction with Antibodies (day 2)

Materials

TRIS buffered saline (TBS)

TRIS buffered saline + Tween 20 (TBST)

Mouse monoclonal anti-human LDH-H subunit

Goat anti mouse IgG-AP conjugate

Procedures

1. Wash the nitrocellulose membrane a few times with distilled water and then with TBST. Pour off the last washing of TBST.

2. Place the nitrocellulose membrane into a tray with 15 mL of diluted primary antibody in TBST. The antibody was diluted 1:5000 in TBST. Incubate the tray for 30 min at room temperature on a shaker.

3. Wash the membrane three times for 10 min each in 20 mL of TBST to remove unbound antibody.

4. Replace the TBST with 15 mL of diluted secondary antibody. The goat anti-mouse IgG-AP conjugate was diluted 1:2500 in TBST.

5. Incubate at room temperature for 30 min as before.

6. Pour off the secondary Ab-AP conjugate and wash the membrane three times for 10 min each with 20 mL of TBST.

7. Blot the membrane damp dry on filter paper and transfer it to 10 mL of the color development solution. Reactive areas will turn purple in 15 to 30 min.

8. When the blot has developed to a desired intensity, stop the reaction by rinsing the membrane with distilled water. The membrane can be air dried for storage on filter paper. Protect it from light.

Analysis of Results

Experiment 10a: Western Blot of LDH

Data

Draw a picture of your gel and blot showing the location of the bands and the identification of the lanes. Use a separate sheet of paper or turn in a photograph if you have one.

Calculations and Questions

1. Which of the bands on the native gel represents LDH?

2. Which LDH isozymes show up on the blot?

3. What can you say about the specificity of the antibody to human LDH-H?

4. Suppose that upon developing your blot, you see more than the bands you expect. What would be some possible reasons for this?

5. Give the mechanism/purpose of the following steps in the blotting procedure. Why did you do it and/or how does it work?
    a. Tagging the secondary antibody with alkaline phosphatase.

    b. Soaking in milk solution.

    c. Running a native acrylamide gel.

6. How could you test the efficiency of transfer of the proteins from the gel to the nitro-cellulose membrane?

Additional Problem Set

1. Why do most Western blot experiments use two antibodies? Why not just tag the primary antibody with horseradish peroxidase?

2. If you were trying to design a Western blot experiment, how would you figure out the time and voltage needed to transfer proteins to the nitrocellulose membrane?

3. When a native acrylamide gel is used for a Western blot, the proteins that ran the farthest down the gel usually transfer to nitrocellulose the fastest. Why is that?

4. Sometimes Western blots are done with native gels and other times with SDS-PAGE. What would influence your choice of gel type?

5. If you color-developed your nitrocellulose after a Western blot and you saw no bands, what tests could you do to figure out what went wrong?

6. If you color-developed your nitrocellulose after a Western blot and you saw too many bands, what tests could you do to figure out what went wrong?

7. What is the difference between a monoclonal antibody and a polyclonal antibody? Why would you choose one or the other as part of a Western blot experiment?

8. Why is it beneficial to use a polyclonal antibody as a secondary, enzyme-labeled antibody for a Western blot experiment?

Webconnections

For a list of Web sites related to the material covered in this chapter, go to **Webconnections** at the *Experiments in Biochemistry* site on the Saunders College Publishing Web page. You can access this page at http://www.saunderscollege.com. **Webconnections** are under *Experiments in Biochemistry* in the Biochemistry portion of the Chemistry page.

References and Further Reading

Bio Rad, *Life Science Research Products Catalog,* 1997.

M. Campbell, *Biochemistry,* Saunders College Publishing, 1998.

D. Chattopadhya, R. K. Aggarwal, U. K. Baveja, V. Doda, and S. Kumari, (1998), *J. Clin. Virol.* **11(1).** Evaluation of Epidemiological and Serological Predictors of Human Immunodeficiency Virus Type-1 Infection among High-Risk Professional Blood Donors with Western Blot Undeterminate Results.

O. Cruz Sui, M. T. Perez Guevara, M. Izquierdo Marquez, L. Lobaina Batelemy, I. Ruibal Brunet, and E. Silva Cabrera, (1997), *Rev. Cubana Med. Trop.* **49(1).** Evaluation of a Western Blot System for the Confirmation of HIV-1 Antibodies.

C. Dartsch and L. Persson, (1998), *Int. J. Biochem Cell Biol.* **30(7).** Recombinant Expression of Rat Histidine Decarboxylase: Generation of Antibodies Useful for Western Blot Analysis.

B. S. Dunbar, *Protein Blotting: A Practical Approach,* IRL Press at Oxford University Press, 1994.

D. Egger and K. Bienz, (1998), *Methods Mol. Biol.* **80.** Colloidal Gold Staining and Immunoprobing on the Same Western Blot.

M. P. Foschini, S. Macchia, L. Losi, A. P. Dei Tos, G. Pasquinelli, L. di Tommaso, S. Del Duca, F. Roncaroli, and P. R. Dal Monte, (1998), *Virchows Arch.* **433(3).** Identification of Mitochondria in Liver Biopsies. A Study by Immunohistochemistry, Immunogold and Western Blot Analysis.

F. J. Frost, A. A. de la Cruz, D. M. Moss, M. Curry, and R. L. Calderon, (1998), *Epidemiol. Infect.* **121(1).** Comparisons of ELISA and Western Blot Assays for Detection of *Cryptosporidium* Antibody.

R. C. Jack, *Basic Biochemical Laboratory Procedures and Computing*, Oxford University Press, 1995.

E. M. Reiche, M. Cavazzana, H. Okamura, E. C. Tagata, S. I. Jankevicius, and J. V. Jankevicius, (1998), *Am. J. Trop. Med. Hyg.* **59(5).** Evaluation of the Western Blot in the Confirmatory Serologic Diagnosis of Chagas' Disease.

J. F. Robyt and B. J. White, *Biochemical Techniques,* Brooks Cole, 1990.

Chapter 11

Restriction Enzymes

Topics

Introduction

This chapter begins our study of molecular biology. All of the modern techniques of molecular biology and our study of DNA were made possible by the discovery of restriction enzymes (nucleases), enzymes that cut DNA at specific sequences. Hundreds of different restriction enzymes have now been isolated and purified from different bacteria.

11.1   Restriction Nucleases

Many types of enzymes modify nucleic acids. One particular type is called a **nuclease,** and it acts to catalyze the hydrolysis of the phosphodiester backbone of nucleic acids. Some nucleases are specific for DNA, others for RNA. Some only cut from an existing end of the nucleic acid. These are called **exonucleases.** Others cut from the inside and are called **endonucleases.** One specific type, called **restriction endonucleases,** has played an important role in the development of modern recombinant DNA technology.

Much of what we know today about molecular biology was made possible by the discovery of restriction endonucleases from bacteria. Bacteria are attacked often by viruses (called bacteriophages) which are dangerous to them. While studying bacteria and the phages that infect them, it was discovered that bacteriophages that grew well in one type

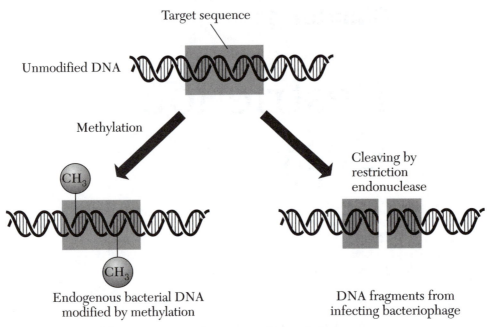

Target sequence

Unmodified DNA

Methylation

Cleaving by
restriction
endonuclease

CH₃

CH₃

Endogenous bacterial DNA
modified by methylation

DNA fragments from
infecting bacteriophage

Figure 11.1 Cleavage of foreign DNA and protection by methylation

of bacteria often grew poorly in another. It was said that their growth was *restricted*. It turned out that there were differences between the DNA of the bacteria and that of the phage based on certain nucleic acid bases that were methylated. The bacteria possessed a type of enzyme that would cleave DNA at specific sites unless the DNA was methylated at those sites, as its own DNA was. Figure 11.1 shows this concept.

For example, the restriction enzyme from *Haemophilus influenzae* Rd, *Hind*III, cuts DNA at

$$5'\text{-A}^{\downarrow}\text{AGCT T-}3'$$
$$3'\text{-T TCGA}_{\uparrow}\text{A-}5'$$

whereas the restriction enzyme from *Escherichia coli* Ry13, *Eco* RI, cuts at

$$5'\text{-G}^{\downarrow}\text{AATT C-}3'$$
$$3'\text{-C TTAA}_{\uparrow}\text{G-}5'$$

Table 11.1 shows some typical restriction nucleases and where they cut. Note that some enzymes, such as *Hae* III, cut straight across from each other leaving what is called a **blunt end,** whereas most cut at different positions leaving what are called **sticky ends.**

11.2 Restriction Maps

A restriction map is a drawing of a piece of DNA showing exactly where the restriction sites are for the enzymes of interest. The DNA is numbered by virtue of its bases, with one end being arbitrarily called zero and the other end being the total number of bases in the piece of DNA. As an analogy, think of a ruler that has a division every millimeter for 300 mm. If we took a knife and cut the ruler at 50 mm and 150 mm, the knife would be the equivalent

Table 11.1 Restriction Endonucleases and Their Cleavage Sites

Enzyme*	Recognition and Cleavage Site
*Bam*HI	5'-G↓GATCC-3' 3'-CCTAG↑G-5'
*Eco*RI	5'-G↓AATTC-3' 3'-CTTAA↑G-5'
Hae III	5'-GG↓CC-3' 3'-CC↑GG-5'
*Hind*III	5'-A↓AGCTT-3' 3'-TTCGA↑A-5'
Hpa II	5'-C↓CGG-3' 3'-GGC↑C-5'
Not I	5'-GC↓GGCCGC-3' 3'-CGCCGG↑CG-5'
Pst I	5'-CTGCA↓G-3' 3'-G↑ACGTC-5'

Arrows indicate the phosphodiester bonds cleaved by the restriction endonucleases.

*The name of the restriction endonuclease consists of a three-letter abbreviation of the bacterial species from which it is derived—for example, *Eco* for *Escherichia coli*.

of a restriction enzyme, and the 50 and 150 mm designations would be the restriction sites. To draw such a restriction map, we would draw something like this:

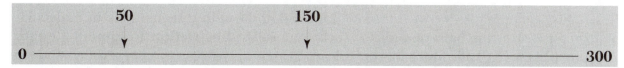

If we then cut the ruler with a pair of scissors at 100 mm, the scissors would be the equivalent of a second restriction enzyme. If we wanted to draw the restriction map of the two, it would look like this:

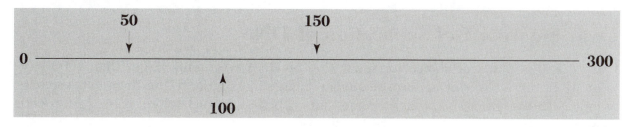

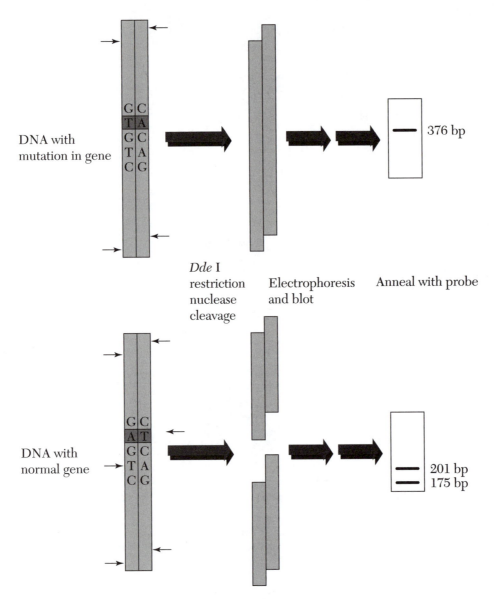

Figure 11.2 Restriction fragment length polymorphism of the β-globin gene

Restriction mapping is one of the most powerful and exciting techniques in molecular biology today. It is the basis of a related technique called **Restriction Fragment Length Polymorphism (RFLP).** Much has been learned about gene structure, especially mutated forms of genes, with RFLP technology. It is also the cornerstone of modern forensic science. If necessary, the final step of paternity testing uses RFLPs. It is also used to prove or disprove the identity of a suspected criminal from blood or other tissue samples.

11.3   Agarose Gel Separation of DNA

Agarose is the matrix of choice for separation of DNA fragments larger than 1000 base pairs. All of the DNA is of uniform shape and charge/mass ratio, so the fragments separate by molecular weight or size, which we measure as base pairs or kilobase pairs. Just as with

➲ When you are plotting the size of DNA fragments vs. migration, don't force the data points into a straight line. You often have a curve.

molecular weight determination of proteins with SDS-PAGE, a plot of log MW of DNA vs. distance migrated should give a straight line. "Should" doesn't always equate to "did," however, so you may have to draw a curve.

The fragments separate on the gel and are compared to a standard. That is the easy part. The hard part is determining what the fragment pattern means. Here is where most students have problems with this type of experiment. Continuing with our ruler analogy, if we cut the ruler with the knife at position 50 mm and 150 mm, we will get fragments that are 50 mm, 100 mm, and 150 mm long. If you cannot immediately see why this is so, do not read any further until you understand it.

If we could separate our ruler pieces on a gel, they would run in order, with the smallest fragment (50 mm) running the fastest, followed by the middle piece (100 mm), and finally the largest piece (150 mm). When we go to plot the fragment size vs. distance migrated, we use the numbers 50, 100, and 150, because that is how long the pieces are. On our gel, we would see three pieces. If we also ran a gel of the pieces we obtained from cutting the ruler with the scissors, we would see two pieces, one corresponding to 100 mm and one corresponding to 200 mm.

To make a complete restriction map, we often have to cut with more than one restriction enzyme at the same time. This is called a double digest. By cutting first with one enzyme, then another, and then both, we get valuable information about where the enzymes cut. In our ruler analogy, if we cut with both the scissors and the knife, we would see pieces of 50 mm and 150 mm. The 50-mm piece we would see would actually represent several different pieces of the same size, but they would show up at the same place on the gel.

Figure 11.3 shows what our ruler analogy gel might look like if we did a traditional experiment designed to figure out where the scissors and the knife cut. To do this type of experiment, we usually know where one of the enzymes cuts. It is the second one we are trying to figure out. We can do this by analyzing the gel from both single digests and the double digest.

11.4   Staining DNA

The most common way to visualize DNA is by putting the chemical ethidium bromide into the gel or reservoir buffer. Ethidium bromide intercalates into DNA and fluoresces bright orange when exposed to UV light. If EtBr is in the buffer during the electrophoresis, then the progress can be monitored and one can actually watch the bands separate.

The disadvantage to this is that *EtBr is a strong carcinogen and must be handled with extreme care.* For this reason, we often let the EtBr soak into the gel after the electrophoresis is over. This way we can keep the ethidium localized to one room. The downside, of course, is that this takes longer.

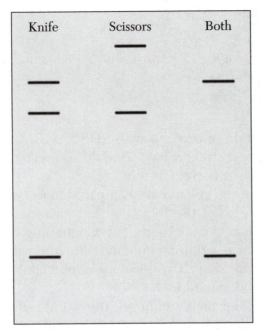

Figure 11.3 The ruler analogy for restriction digests

The extent of the run can be monitored as usual by using tracking dyes. On a 1% agarose gel, bromophenol blue migrates with fragments of around 200 base pairs whereas xylene cyanol migrates with fragments closer to 3000 base pairs.

11.5 Phage λ DNA

Bacteriophage λ is a virus that infects *E. coli* and is one of the most studied phages. The entire phage is 48,502 base pairs in length and contains restriction sites for many endonucleases. *Hind*III cuts λ DNA seven times, at the positions shown in Table 11.2.

A couple of things to note are that the fragments generated are based on the difference between two adjacent cuts or between a cut and either end. Also, the smallest fragment rarely shows up on a gel. The 564–base pair fragment usually does show up if the gel isn't run too long.

Table 11.2 *Hind*III Digestion of λ DNA

Cut Sites from Zero Point	Fragment Generated from Zero or From Last Cut (kb)
23,130	23.1
25,157	2.0
27,479	2.3
36,895	9.4
37,459	0.564
37,584	0.125
44,141	6.5
48,502 (end, not a site)	4.3

A restriction map showing the *Hin*dIII sites on λ DNA would look like the following:

```
                                          37.6
                     25.1                 37.4
             23.1    27.4                 36.9           44.2
   23.1              ▼2 ▼2.3▼    9.4     ▼▼▼   6.5   ▼ 4.3
0 ─────────────────────────────────────────────────────────── 48.5
```

When λ DNA is cut with *Hin*dIII, the resulting banding pattern looks like Figure 11.4. This is a very common set of DNA size markers. You can either create these markers yourself, or buy commercially prepared λ DNA/*Hin*dIII fragments.

⟜ Practice Session 11.1

If you cut λ DNA with *Hin*dIII and a second enzyme that cuts one time at position 33498, what would the resulting gel look like?

Looking at the λ map shown earlier, we can see that the second enzyme cuts in between two of the *Hin*dIII sites, namely the ones that give the 9.4-kb (kilobase) piece. The 9.4-kb piece is the second band on the gel shown in Figure 11.5. If we cut the λ DNA with just the second enzyme, it would cleave the DNA into two pieces. One would be 33,498 base pairs long. This would be a big piece that would be high up on the gel. The second piece would be 48,502 − 33,498 = 15,004 base pairs.

This latter piece would show up lower than the first band from the *Hin*dIII (the 23.1-kb piece) but higher than the 9.4-Kb piece. If we ran a double digest, the majority of the bands from the *Hin*dIII digestion would be untouched, because the second enzyme does not cleave in their regions. The only band that would change would be the 9.4-kb fragment. It would be cut by the second enzyme and would not show up in the double-digest lane. In place of the 9.4 Kb fragment, two fragments would be generated. The first would

Figure 11.4 λ DNA cut with *Hin*dIII

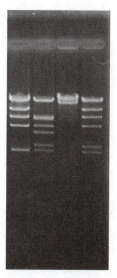

Figure 11.5 Figure 11.6 λ DNA cut with *Hin*dIII (Lanes 1 and 4), a second enzyme (Lane 3), and both *Hin*dIII and the second enzyme (Lane 2)

be 33,498 − 27,479 = 6019 base pairs. The second would be 36,895 − 33,498 = 3397 base pairs. Figure 11.5 shows what such a gel would look like.

11.6   Why Is This Important?

Molecular biology is *the* field these days. It is where the money is, both commercially and academically. All we know about DNA and molecular biology came from the discovery of restriction enzymes. It seems so simple. Just cut up some DNA! It is such a powerful tool given the diverse nature of the many restriction enzymes. Besides classical basic research, there are many fascinating ways that restriction enzymes are used. Forensic medicine uses restriction digestion of samples to identify people. Many criminals have been caught and identified by the specific restriction digest pattern of their DNA.

ESSENTIAL INFORMATION

Restriction enzymes are used to cut DNA at very specific sequences. They are the backbone of the whole field called molecular biology. When DNA is cut with restriction enzymes, fragments result that may be separated on an agarose gel. With the agarose gel, we can measure the size of the fragments if we have standards to compare them to. A common DNA standard is generated by cutting phage λ DNA with *Hind*III. By cutting DNA with a second enzyme and comparing to the first, we can figure out where on the DNA the second enzyme cuts. Eventually, we can make a restriction map, which shows the exact location of restriction enzyme cut sites on a piece of DNA.

Experiment 11

Analysis of DNA Restriction Fragments

In this experiment, you will use restriction enzymes to digest phage λ DNA and then run an agarose gel to separate the fragments. Using information from a known restriction enzyme, you will be able to determine where on the phage λ DNA a second, unknown enzyme cuts.

Pre Lab Questions

1. What will you do immediately before having the restriction enzymes added to your reactions?

2. What safety techniques are we using with respect to use of ethidium bromide?

Objectives:

Upon successful completion of this experiment, the student will be able to

1. Prepare agarose gels and load DNA samples.

2. Conduct restriction digestion reactions.

3. Stain gels with ethidium bromide to visualize DNA.

4. Analyze DNA fragments and determine the molecular weight of the fragments.

5. Draw a partial restriction enzyme map.

In this experiment you will digest λ DNA with *Hind*III, *Mys* I (a mystery endonuclease), and a mixture of both to see the restriction map produced.

Experimental Procedures

Materials

Phage λ DNA and commercial λ DNA/*Hind*III fragments

*Hind*III

Mys I

Reaction buffer for endonucleases

TAE buffer

37° and 65°C water bath

Electrophoresis equipment

DNA stain (EtBr will be added at the end)

Pipetmen and tips

Methods

Digestion of λ DNA

1. From the stockroom you will receive microfuge tubes with λ DNA, reaction buffer, and commercial *Hind*III λ fragments.

2. Get an ice bucket to keep all of your samples in when not incubating at a prescribed temperature.

3. Set up 3 reactions as shown, except for the *Hind*III and *Mys* I.

Reaction	1	2	3
10× reaction buffer	1 μL	1 μL	1 μL
dd water	6 μL	6 μL	5 μL
λ DNA (0.5 μg/(μL)	2 μL	2 μL	2 μL
*Hind*III	1 μL	—	1 μL
Mys I	—	1 μL	1 μL

4. Spin the tubes in the microfuge for a few seconds. Take them to the instructor or TA to have the enzymes added. Mix and spin again.

5. Incubate the tubes at 37°C for 20 min.

> **Caution!**
> Anytime you are anywhere near ethidium bromide, you must be properly dressed. This includes gloves, long sleeves, and eye protection.

6. When the incubation is over, add 2 μL of bromophenol blue/xylene cyanol tracking dye to the three samples and to a tube of commercial λ fragments.

7. Place the four tubes in the 65°C waterbath for 5 min and then return them to the ice bucket quickly.

Agarose Gel Electrophoresis

1. While the digestion is proceeding, make 40 mL of 1% agarose in TAE buffer for the Fisher units (see Experiment 9).

2. Level a casting tray, pour the gel, insert the comb, and let the gel harden for 15 min.

3. Carefully remove the rubber stopper and comb.

4. Pour the TAE reservoir buffer so that it is just over the gel. Make sure the + pole is away from the wells.

5. Add your samples to the wells.

6. Set the voltage to 120.

7. When you are done electrophoresing, place the gel into a *clean* plastic container, taped and labeled. Take your gel into the predetermined room to have the ethidium bromide added.

8. View the gel and have a picture taken.

9. Measure the migration of the bands on the picture.

Analysis of Results

Experiment 11: Restriction Fragment Analysis

Data

1. Turn in your photograph or Xerox of the gel, labeling the lanes. If neither is available, draw a picture of the gel on a separate sheet of paper.

2. Measure the distances of the bands starting with the bands the highest up (closest to the wells).

Lane 1	Lane 2	Lane 3	Lane 4
_____	_____	_____	_____
_____	_____	_____	_____
_____	_____	_____	_____
_____	_____	_____	_____
_____	_____	_____	_____
_____	_____	_____	_____
_____	_____	_____	_____
_____	_____	_____	_____
_____	_____	_____	_____

Calculations

1. Graph the log base pairs vs. migration distance for the *Hind*III fragments. Draw a line or smooth curve that really fits the data. If the points do not make a straight line, do not force them. Determine the base pairs for the unique *Mys* I fragments and double-digest fragments.

2. What fragments from one digest have a restriction site for the other endonuclease within them? (For example, the 6.5-kb-*Hind*III fragment has a Mys site within it. . . .)

3. Using the known location of the *Hind*III sites and their fragment sizes, the calculated sizes of the *Mys* I fragments, and the information from the double digest, make a restriction map showing as many of the *Mys* I sites as you can. It is easiest if you include the *Hind*III sites on the top and the *Mys* I sites on the bottom of a line. If your results are not good, get a picture of a more appropriate gel that you can use for this question.

Questions

1. The restriction enzyme *Bst* XI recognizes the sequence CCANTGG whereas *Hha* I recognizes GCGC. Which of these two enzymes is more likely to cut a given piece of DNA most often? Explain.

2. A DNA fragment of 3300 bp was digested with *Eco*RI alone, with *Hind*III alone, and with *Eco*RI and III together. The fragments obtained and their sizes are shown in the figure. Draw a restriction map based on this information.

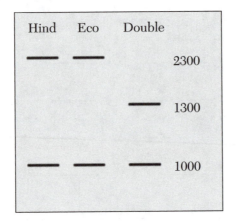

Additional Problem Set

1. What is the importance of methylation in the activity of restriction endonucleases?

2. Why do restriction endonucleases not hydrolyze DNA from the organism that produces them?

3. What is a palindrome? What English words or sentences are palindromes?

4. What is the importance of "sticky ends" in molecular biology?

5. The restriction enzyme *Xho* I cuts phage λ DNA at position 33,498. Draw a picture of the gel you would see if *Xho* I were used as the mystery enzyme in Experiment 11.

6. The restriction enzyme *Sst* I cuts phage λ DNA at positions 24,472 and 25,877. Draw a picture of the gel you would see if *Sst* I were used as the mystery enzyme in Experiment 11.

7. The restriction enzyme *Xba* I cuts phage λ DNA at position 24,508. Draw a picture of the gel you would see if *Xba* I were used as the mystery enzyme in Experiment 11.

8. For which of the two restriction enzymes *Xho* I or *Sst* I would it be best to run a 1.5% agarose gel instead of a 1% gel? (See the information in problems 5 and 6.)

Webconnections

For a list of Web sites related to the material covered in this chapter, go to **Webconnections** at the *Experiments in Biochemistry* site on the Saunders College Publishing Web page. You can access this page at http://www.saunderscollege.com. **Webconnections** are under *Experiments in Biochemistry* in the Biochemistry portion of the Chemistry page.

References and Further Reading

J. N. Anderson, *A Laboratory Course in Molecular Biology*, Modern Biology, 1986.

F. Ausubel, R. Brent, R. Kingston, D. Moore, J. Seidman, J. Smith, and K. Struhl, *Short Protocols in Molecular Biology*, Wiley, 1992.

P. Berg and M. Singer, *Dealing with Genes: The Language of Heredity*, University Science Books, 1992.

W. Bodmer and R. McKie, *The Book of Man: the Human Genome Project and the Quest to Discover our Genetic Heritage*, Simon and Schuster, 1995.

R. F. Boyer, *Modern Experimental Biochemistry*, Addison-Wesley, 1993.

M. Campbell, *Biochemistry*, Saunders College Publishing, 1998.

R. L. Dryer and G. F. Lata, *Experimental Biochemistry*, Oxford University Press, 1989.

W. A. Haseltine, (1997), *Sci. Am.* **276(3).** Discovering Genes for New Medicines.

A. L. Lehninger, D. L. Nelson, and M. M. Cox, *Principles of Biochemistry*, Worth, 1993.

E. Pennisi, (1996), *Science,* **273.** Chemical Shackles for Genes?

L. Roberts, (1990), *Science,* **249.** New Scissors for Cutting Chromosomes.

J. F. Robyt and B. J. White, *Biochemical Techniques*, Brooks Cole, 1990.

L. Stryer, *Biochemistry*, 3rd ed., Freeman, 1988.

Chapter 12

Cloning and Expression of Foreign Proteins

Topics

Introduction

In this chapter we discuss the methods used to recombine DNA from different sources, a procedure known as cloning. This is usually done with the ultimate goal of producing large quantities of a protein of interest, such as an important hormone or immunity factor. With cloning, the rapid proliferation of a bacterial strain can be used to produce much larger quantities of a mammalian protein than would be possible by traditional purification means.

12.1   Recombinant DNA

The entire field of molecular biology grew from the ability to manipulate genetic information using recombinant DNA technology. Recombinant DNA is DNA from two different sources that have

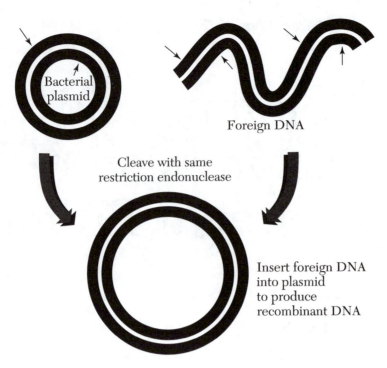

Bacterial plasmid

Foreign DNA

Cleave with same
restriction endonuclease

Insert foreign DNA
into plasmid
to produce
recombinant DNA

Figure 12.1   Creating recombinant DNA

been recombined so that a gene of interest can be studied. Recombinant DNA allows researchers to clone individual genes so they can study both their structure and function separate from the organism in which they were originally isolated. This chapter explains the process of cloning and expressing foreign proteins in bacteria. Figure 12.1 shows the basics of creating recombinant DNA.

Insulin was the first commercially available, genetically engineered product that was a direct result of expressing a foreign protein in bacteria. Since then, many products have been created using this technology, including some cancer-fighting drugs such as interferon. The term *cloning* has many different meanings depending on the context in which it is used. Molecular biologists use the term to describe the process of creating recombinant DNA molecules. In microbiology, you learn that bacteria replicate by binary fission, giving rise to a colony of cells that are identical clones. A stricter definition of a clone is a group of genetically identical organisms. Perhaps the most common example of a clone is an identical twin.

Here is some terminology that you should become familiar with as you read through this chapter. The terms *foreign DNA* and *insert* describe the segment of DNA, usually a single gene, that is of interest to the researcher. This is the segment of DNA that is usually being cloned. This segment of DNA contains a gene or at least part of a gene. Most of the time, the segment of DNA encodes a protein and is referred to as the **open reading frame (ORF).** To clone the segment of DNA and express it in bacteria you must first put it into a carrier molecule or **vector.** A vector is piece of DNA that allows replication of the foreign DNA. Common vectors used in molecular biology are plasmids and phages. We will only deal with plasmids in this chapter. Of course, recombinant DNA technology would not be possible without the discovery of **restriction endonucleases,** which we studied in Chapter 11. Another important enzyme is **DNA ligase,** which catalyzes the resealing of the DNA backbone once it has been cut.

Once the foreign DNA or insert has been ligated or cloned into a vector, it is transferred to a host cell line via a process referred to as **transformation.** The host cell then replicates, producing identical offspring (clones) as well as replicating the foreign gene. Thus, as long as the foreign gene can be replicated, we can clone the gene for as long as the bacteria continue to clone themselves. This leads to a huge amplification of the cloned DNA, because most bacteria divide every 20 min under optimal conditions.

Bacteria are very prolific, so you need to be able to find the bacterial cells that actually were transformed. This process is called **selection.** The vector will have some type of marker that allows you to spot the bacterial cells that took up the vector/insert. Common selectable markers are genes for resistance to antibiotics.

Once you have large quantities of the cloned gene, you can have the gene expressed. This means the gene is transcribed to messenger RNA (mRNA), and then the mRNA is translated into protein. This is the process of **expression** and often requires a specific vector and bacterial cell line that have the ability to express foreign proteins.

12.2   Vectors

A vector is a piece of DNA that can be used to carry foreign DNA into host cells. A vector, by definition, replicates autonomously (separately from chromosomal DNA) and can therefore be used as a cloning vehicle for identifying newly created recombinant DNA molecules. Once the foreign DNA is recombined with a vector, it is transferred to a host cell. The vector with the foreign DNA is replicated, producing clones that can be isolated and analyzed. The most common vectors are plasmids, bacteriophages, and cosmids. All can be used for cloning, but we will only consider plasmids in this chapter.

Plasmids are small autonomously replicating pieces of DNA that were originally identified in bacteria because they contain genes that allow for resistant to certain antibiotics. It was then found that bacteria could pass around these antibiotic resistant plasmids from cell to cell. The cell receiving the plasmid would then become resistant to the antibiotic and could grow in high concentrations of the antibiotic. The plasmid's ability to confer antibiotic resistance is necessary for use in cloning, because researchers are only interested in cells that contain the plasmid. However, one unfortunate consequence of plasmid transfer between bacteria is that more and more bacteria are becoming resistant to certain antibiotics. Figure 12.2 shows a typical plasmid.

Plasmids have been modified to allow their use in cloning. Today's plasmids have all been designed with the researcher's needs in mind and do not resemble the original bacterial plasmids from which they were derived. A plasmid must have several features in order to be used as a vector for cloning. For the plasmid to be replicated independently of the host genome, it must have its own **origin of replication.** This is usually indicated on a plasmid map as **oriC.** A plasmid must have a **selectable marker,** such as gene that confers resistance to an antibiotic. The presence of a selectable marker allows for growth of *Escherichia coli* in the presence of an antibiotic, thus allowing selection of cells that contain the plasmid. The plasmid map shown in Figure 12.2 has one selectable marker, the gene for ampicillin resistance Amp^r. The plasmid must also have many unique restriction sites for use when inserting the foreign piece of DNA into the plasmid. These unique sites are often concentrated on a region of high restriction site density called a **multiple cloning site (MCS).** With the

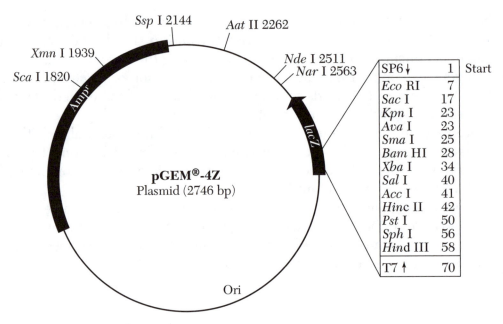

SP6 ↓	1	Start
Eco RI	7	
Sac I	17	
Kpn I	23	
Ava I	23	
Sma I	25	
Bam HI	28	
Xba I	34	
Sal I	40	
Acc I	41	
Hinc II	42	
Pst I	50	
Sph I	56	
Hind III	58	
T7 ↑	70	

Figure 12.2 pGEM plasmid vector

pGEM® vector shown in Figure 12.2, the MCS is within the *lac*Z gene and has 13 different restriction enzyme sites in 58 base pairs.

The purpose of having the *lac*Z gene in the MCS is that it allows for easy screening (**blue/white screening**) of the transformed cells that have the foreign DNA. When the MCS is within the *lac*Z gene, insertion of foreign DNA will disrupt the gene, rendering it nonfunctional. If for some reason the foreign DNA is not inserted into the MCS, the host cell will then produce the β-galactosidase protein from the *lac*Z gene. In order to screen for insertion of foreign DNA when cloning, an *E. coli* strain that contains a nonfunctional β-galactosidase protein is used. The recombinant plasmid is transformed into this strain, and the resulting transformants are grown on plates containing a dye called X-gal, which turns blue when the β-galactosidase protein is present. Thus, if an *E. coli* cell contains a vector with foreign DNA in the MCS, it will remain white. If the vector does not contain the insert, the cells will turn blue because they are producing the β-galactosidase protein. This allows a quick distinction between cells that took up a plasmid with the insert and those that took up a plasmid without the insert.

12.3   Foreign DNA

The term *foreign DNA* refers to the sequence of DNA that the researcher is interested in. Molecular biologists often refer to this DNA as the "insert" when cloning. Most of the time, the foreign DNA is a gene encoding a single polypeptide, but this is not always the case. This foreign DNA can be obtained using any number of standard techniques (e.g., PCR). Once the DNA is cloned into a vector, it can then be used for many different applications. For example, it can be used to express and purify the protein in *E. coli*. If the gene encodes an unknown protein, then the researcher can sequence the DNA and deduce (using the

```
  1 GGAAAAGCTGCACTCAGGAGGAGACACATTCCAGCTCGCGCCTTCGCCTTTATCTTAAAAACACCTCCCTCCCAAAAAAAAGAAAAAAAAACCTCAAG 98

 99 ATG TCC ACC AAG GAG AAG CTC ATC GAC CAC GTG ATG AAG GAG GAG CCT ATT GGC AGC AGG AAC AAG GTG ACG GTG 173
  1  M   S   T   K   E   K   L   I   D   H   V   M   K   E   E   P   I   G   S   R   N   K   V   T   V   25

174 GTG GGC GTT GGC ATG GTG GGC ATG GCC TCC GCC GTC AGC ATC CTG CTC AAG GAC CTG TGT GAC GAG CTG GCC CTG 248
 26  V   G   V   G   M   V   G   M   A   S   A   V   S   I   L   L   K   D   L   C   D   E   L   A   L   50

249 GTT GAC GTG ATG GAG GAC AAG CTG AAG GGC GAG GTC ATG GAC CTG CAG CAC GGA GGC CTC TTC CTC AAG ACG CAC 323
 51  V   D   V   M   E   D   K   L   K   G   E   V   M   D   L   Q   H   G   G   L   F   L   K   T   H   75

324 AAG ATT GTT GGC GAC AAA GAC TAC AGT GTC ACA GCC AAC TCC AGG GTG GTG GTG GTG ACC GCC GGC GCC CGC CAG 398
 76  K   I   V   G   D   K   D   Y   S   V   T   A   N   S   R   V   V   V   V   T   A   G   A   R   Q  100

399 CAG GAG GGC GAG AGC CGT CTC AAC CTG GTG CAG CGC AAC GTC AAC ATC TTC AAG TTC ATC ATC CCC AAC ATC GTC 473
101  Q   E   G   E   S   R   L   N   L   V   Q   R   N   V   N   I   F   K   F   I   I   P   N   I   V  125

474 AAG TAC AGC CCC AAC TGC ATC CTG ATG GTG GTC TCC AAC CCA GTG GAC ATC CTG ACC TAC GTG GCC TGG AAG CTG 548
126  K   Y   S   P   N   C   I   L   M   V   V   S   N   P   V   D   I   L   T   Y   V   A   W   K   L  150

549 AGC GGG TTC CCC CGC CAC CGC GTC ATC GGC TCT GGC ACC AAC CTG GAC TCT GCC CGT TTC CGC CAC ATC ATG GGA 623
151  S   G   F   P   R   H   R   V   I   G   S   G   T   N   L   D   S   A   R   F   R   H   I   M   G  175

624 GAG AAG CTC CAC CTC CAC CCT TCC AGC TGC CAC GGC TGG ATC GTC GGA GAG CAC GGA GAC TCC AGT GTG CCT GTG 698
176  E   K   L   H   L   H   P   S   S   C   H   G   W   I   V   G   E   H   G   D   S   S   V   P   V  200

699 TGG AGT GGA GTG AAC GTT GCT GGA GTT TCT CTG CAG ACC CTT AAC CCA AAG ATG GGG GCT GAG GGT GAC ACG GAG 773
201  W   S   G   V   N   V   A   G   V   S   L   Q   T   L   N   P   K   M   G   A   E   G   D   T   E  225

774 AAC TGG AAG GCG GTT CAT AAG ATG GTG GTT GAT GGA GCC TAC GAG GTG ATC AAG CTG AAG GGC TAC ACT TCC TGG 848
226  N   W   K   A   V   H   K   M   V   V   D   G   A   Y   E   V   I   K   L   K   G   Y   T   S   W  250

849 GCC ATC GGC ATG TCC GTG GCT GAC CTG GTG GAG AGC ATC GTG AAG AAC CTG CAC AAA GTG CAC CCA GTG TCC ACA 923
251  A   I   G   M   S   V   A   D   L   V   E   S   I   V   K   N   L   H   K   V   H   P   V   S   T  275

924 CTG GTC AAG GGC ATG CAC GGA GTA AAG GAC GAG GTC TTC CTG AGT GTC CCT TGC GTC CTG GGC AAC AGC GGC CTG 998
276  L   V   K   G   M   H   G   V   K   D   E   V   F   L   S   V   P   C   V   L   G   N   S   G   L  300

 999 ACG GAC GTC ATT CAC ATG ACG CTG AAG CCC GAA GAG GAG AAG CAG CTG GTG AAG AGC GCC GAG ACC CTG TGG GGC 1073
 301  T   D   V   I   H   M   T   L   K   P   E   E   E   K   Q   L   V   K   S   A   E   T   L   W   G  325

1074 GTA CAG AAG GAG CTC ACC CTG TGA GTGTCGCTCCTCTGATTTCTCCAGTCCGCCCTGAAAACACACCAAACACTGTGTGGTTATCCCCTCCC 1165
 326  V   Q   K   E   L   T   L   *                                                                       333
```

Figure 12.3 Barracuda LDH-A gene sequence printout by DNA Strider computer program

genetic code) the amino acid sequence of the protein. This type of analysis can be very useful when trying to understand the function of the protein. The sequence can also be analyzed using computer programs. These programs can search for restriction sites, open reading frames, percentage homology to other genes, and so on. Figure 12.3 shows a printout from a DNA analysis program (DNA Strider™ 1.2). The sequence of barracuda lactate dehydrogenase (LDH) is listed. In this figure the program was instructed to find the first start codon and begin translating it until it encounters a stop codon. The translated amino acid sequence is shown using the single-letter codes for the amino acids.

12.4   Restriction Enzymes

Figure 12.4 gives an example of how a DNA analysis program can analyze DNA sequences for **restriction sites.** A restriction site is a sequence of DNA that is recognized by a restriction endonuclease. These sequences are usually palindromic. Figure 12.5 shows a complete

Positions of Restriction Endonucleases sites (unique sites italicized)

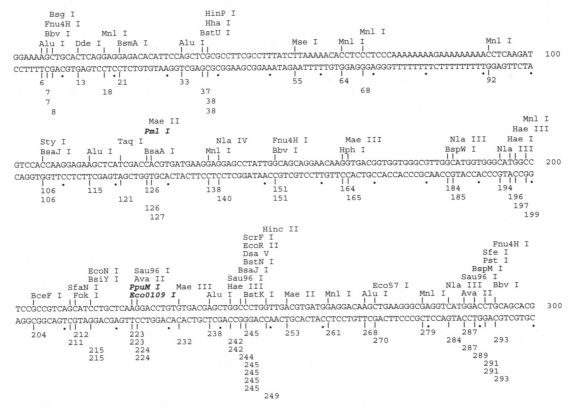

Figure 12.4 Partial restriction map of barracuda LDH-A from the DNA Strider program

restriction map indicating which enzymes cut the barracuda LDH ORF, what the sites are, and where they cut. Another useful piece of information is what enzymes don't cut the DNA. This can be useful when trying to clone the segment of DNA. The DNA Strider program also gives the list of restriction enzymes not found in the DNA, as is shown in Figure 12.6.

Once you have all the information on restriction enzymes and the sequences of your foreign DNA and vectors, you can plan your cloning strategy. To clone the foreign DNA into a suitable vector, you need to find unique restriction sites that are outside the actual gene itself (ORF). If there are no convenient sites, then you can mutate or change the DNA sequence to introduce one. Remember this won't affect the protein because you won't change the ORF. For example, if you wanted to clone the DNA shown in Figure 12.3 and 12.4, you would not want to pick *Alu* I to cut out the gene. Although *Alu* I has a site at position 33, which is upstream of the translation start site (the ATG at position 99), there is also an *Alu* I site at position 115. Thus, if you chose that restriction enzyme, you would cut the coding region of the gene into pieces.

You would instead choose another restriction enzyme that does not cut into the coding region. You would also need to find the same restriction site in the vector you were planing to clone the foreign DNA into. It would be difficult to clone this DNA into the pGEM vector shown in Figure 12.2 because there are no common restriction sites available between

Enzyme		Site		Use	Site position	(Fragment length)	Fragment order
Alw	I	ggatc	4/5	1	1(660) 2	661(1649)	1
AlwN	I	cagnnn/ctg		1	1(1378) 1	1379(931)	2
Ase	I	at/taat		1	1(2175) 1	2176(134)	2
Avr	II	c/ctagg		1	1(1991) 1	1992(318)	2
Ban	II	grgcy/c		1	1(1082) 2	1083(1227)	1
Bbe	I	ggcgc/c		1	1(386) 2	387(1923)	1
Bbs	I	gaagac	2/6	1	1(956) 2	957(1353)	1
Bcl	I	t/gatca		1	1(819) 2	820(1490)	1
Bgl	II	a/gatct		1	1(2161) 1	2162(148)	2
BstE	II	g/gtnacc		1	1(376) 2	377(1933)	1
BstY	I	r/gatcy		1	1(2161) 1	2162(148)	2
Bsu36	I	cc/tnagg		1	1(1204) 1	1205(1105)	2
Drd	I	gacnnnn/nngtc		1	1(341) 2	342(1968)	1
Eae	I	y/ggccr		1	1(1728) 1	1729(581)	2
Ecl136	I	gag/ctc		1	1(1082) 2	1083(1227)	1
Eco0109	I	rg/gnccy		1	1(222) 2	223(2087)	1
Ehe	I	ggc/gcc		1	1(386) 2	387(1923)	1
Esp	I	gc/tnagc		1	1(544) 2	545(1765)	1
Fsp	I	tgc/gca		1	1(2046) 1	2047(263)	2
Gdi	II	yggccg	-5/-1	1	1(1728) 1	1729(581)	2
Kas	I	g/gcgcc		1	1(386) 2	387(1923)	1
Nae	I	gcc/ggc		1	1(383) 2	384(1926)	1
Nar	I	gg/cgcc		1	1(386) 2	387(1923)	1
Nco	I	c/catgg		1	1(1350) 1	1351(959)	2
PflM	I	ccannnn/ntgg		1	1(610) 2	611(1699)	1
Pml	I	cac/gtg		1	1(125) 2	126(2184)	1
PpuM	I	rg/gwccy		1	1(222) 2	223(2087)	1
PshA	I	gacnn/nngtc		1	1(997) 2	998(1312)	1
Sac	I	gagct/c		1	1(1082) 2	1083(1227)	1
SgrA	I	cr/ccggyg		1	1(382) 2	383(1927)	1
Spe	I	a/ctagt		1	1(2299) 1	2300(10)	2
Sph	I	gcatg/c		1	1(933) 2	934(1376)	1
Ssp	I	aat/att		1	1(1359) 1	1360(950)	2
Stu	I	agg/cct		1	1(301) 2	302(2008)	1
Tth111	I	gacn/nngtc		1	1(950) 2	951(1359)	1

Figure 12.5 List of restriction enzymes and sites for barracuda LDH-A from the DNA Strider program

the vector and the upstream noncoding region. Chapter 13 will go into how you might get around this problem.

Another consideration is how many restriction enzymes to use. If you had a DNA insert that was cut out with *Eco*RI at both ends, it would be easy to ligate this into a vector that was opened up with the same enzyme (see Figure 12.1). An important technical consideration is how to keep the vector from simply resealing, because the *Eco*RI leaves sticky ends that are self-complementary. This can be done by removing the phosphate group on the end of the DNA with an enzyme called a phosphatase. As the name implies, it is a specially suited to remove phosphate from DNA. On the other hand, if you cut the foreign DNA with two different enzymes and cut the vector with the same enzymes, you would be able to ligate the foreign DNA into the vector without worrying about the vector reclosing, because the sticky ends would not match.

No Sites found for the following Restriction Endonucleases

Acc	I	gt/mkac	BspE	I	t/ccgga	Mlu	I	a/cgcgt	Sap	I	gcttcttc	1/4
Afl	II	c/ttaag	BspH	I	t/catga	Msc	I	tgg/cca	Sca	I	agt/act
Afl	III	a/crygt	BssH	II	g/cgcgc	Nci	I	cc/sgg	Sfi	I	ggccnnnn/nggcc
Age	I	a/ccggt	BstB	I	tt/cgaa	Nde	I	ca/tatg	Sma	I	ccc/ggg
Apa	I	gggcc/c	Cla	I	at/cgat	Nhe	I	g/ctagc	Sna	I	gta/tac
Asp718		g/gtacc	Eag	I	c/ggccg	Not	I	gc/ggccgc	SnaB	I	tac/gta
Ava	I	c/ycgrg	Eco47	III	agc/gct	Nru	I	tcg/cga	Spl	I	c/gtacg
BamH	I	g/gatcc	EcoR	I	g/aattc	Nsi	I	atgca/t	Sse8337	I	cctgca/gg
Bcn	I	ccs/gg	EcoR	V	gat/atc	Pac	I	ttaat/taa	Swa	I	attt/aaat
Bgl	I	gccnnnn/nggc	Fse	I	ggccgg/cc	PaeR7	I	c/tcgag	Xba	I	t/ctaga
BsaB	I	gatnn/nnatc	Hind	III	a/agctt	Pvu	I	cgat/cg	Xca	I	gta/tac
BsiE	I	cgry/cg	Hpa	I	gtt/aac	Rsr	II	cg/gwccg	Xho	I	c/tcgag
Bsm	I	gaatgc	1/-1	Kpn	I	ggtac/c	Sac	II	ccgc/gg	Xma	I	c/ccggg
Bsp120	I	g/ggccc	Mcr	I	c/grycg	Sal	I	g/tcgac	Xmn	I	gaann/nnttc

Figure 12.6 Restriction sites not found in barracuda LDH-A by the DNA Strider program

⟱▷ **Practice Session 12.1**

If you had the gene shown here and wanted to clone it into the pGEM vector, what restriction enzymes would you choose? The restriction map of an example gene is shown below. The start codon (ATG) is at position 99. The stop codon (TGA) is at position 1095.

Restriction Enzyme Usage

Enzyme	Position	Enzyme	Position
Aat II	30, 2000	Not I	986
Bal I	100	Pst I	1050
BamHI	45	Sac I	70
Dra I	1000	Sal I	500
EcoRI	300, 762	Sca I	35
HindIII	1098	Xba I	150
Kpn I	200	Xho I	684
Nae I	895	Xmn I	467

The strategy is to find two restriction sites that appear inside the multiple cloning site of the pGEM vector and that surround the coding region of the gene. This can be done by trial and error, looking at each available restriction enzyme. For example, what if we wanted to use *Aat* II? That enzyme cuts at position 30, which is before the start codon at position 99. *Aat* II does have a site on the vector at position 2262, but that is outside of the multiple cloning site. That would be a bad choice if you wanted to use the *lacZ* gene for screening. *Sca* I cuts at position 35 on the gene, but inside the ampicillin resistance gene on the plasmid. *Sac* I cuts at position 70 on the gene, which is before the start codon. It also cuts inside the multiple cloning site of the vector. On the other end, we need a restriction enzyme that cuts after the stop codon. There are only two restriction enzymes listed that cut after the stop codon at 1095, *Hin*d III and *Aat* II, the latter of which we have already eliminated. *Hin*dIII does have a site inside the MCS. Thus, the best two enzymes to use would be Sac I and *Hin*dIII.

ESSENTIAL INFORMATION

Cloning involves cutting desired sequences of DNA with restriction enzymes and then placing the DNA into a vector. The best way is to cut the target DNA with two different restriction enzymes and then cut the vector with the same two enzymes. In this way, it becomes much more likely that the target DNA will be incorporated into the vector because a vector cut with two different enzymes is not likely to close back upon itself without the target DNA insert. DNA sequence programs exist that allow you to easily see which restrictions enzymes have sites within your target gene and which do not. Using this information and the similar information from the vector, you can decide which enzymes to use to do your cloning.

12.5   Cell Lines

A **cell line** or **strain** is a specific type of cell, either prokaryotic or eukaryotic, with a defined genotype. When using recombinant DNA technology to clone and express foreign proteins, two types of bacterial cell lines are used—those that are used for cloning foreign DNA and those that are used for expressing foreign proteins.

Each type of cell line has been modified to perform a certain function. In addition, most cell lines are designed to use special vectors that allow them to perform that function. For example, the bacterial strain BL21 (Novagen) is designed for the sole purpose of expressing foreign proteins. This strain contains a chromosomal copy of the gene for T7 RNA Polymerase. This gene is under the control of the *lac*UV5 promoter, which means the gene can be turned on or off by the addition of a certain inducer. To express proteins in this strain, you put your foreign DNA into a specially designed vector that has the T7 promoter in front of the MCS. Once you have inserted your DNA into the vector, the plasmid is transformed into the BL21 strain and the protein is expressed by the addition of inducer. The addition of inducer turns on the T7 RNA polymerase gene and causes the production of T7 RNA polymerase, which recognizes and binds the T7 promoter and transcribes the foreign gene. The mRNA produced from this transcription is then translated into protein.

When cloning experiments are done, the type of cell line used for the recovery of the recombinant plasmid is critical. These types of cloning strains have been modified to ensure high yield and efficient recovery of recombinant plasmid. In addition, some have been modified to allow for screening of recombinant plasmid. For example, if you want to be able to do blue/white screening as described in Section 12.2, not only would you need to use a plasmid with the multiple cloning site in the *lac*Z gene, but you would need a host strain that was deficient in the α-subunit of β-galactosidase.

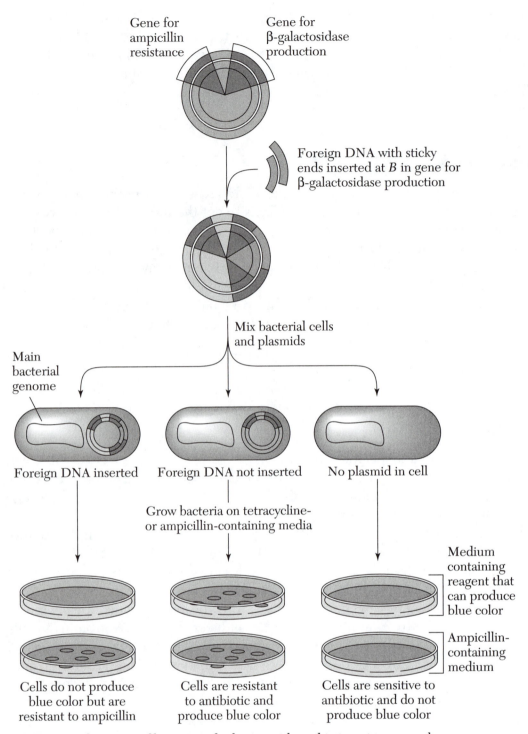

Gene for
ampicillin
resistance

Gene for
β-galactosidase
production

Foreign DNA with sticky
ends inserted at *B* in gene for
β-galactosidase production

Mix bacterial cells
and plasmids

Main
bacterial
genome

Foreign DNA inserted

Foreign DNA not inserted

No plasmid in cell

Grow bacteria on tetracycline-
or ampicillin-containing media

Medium
containing
reagent that
can produce
blue color

Ampicillin-
containing
medium

Cells do not produce
blue color but are
resistant to ampicillin

Cells are resistant
to antibiotic and
produce blue color

Cells are sensitive to
antibiotic and do not
produce blue color

Figure 12.7 Transformation of bacteria and selection with antibiotic resistance markers

12.6   Transformation

Transformation is the process by which there is a heritable change in a cell or an organism brought about by exogenous DNA. In molecular biology, the exogenous DNA is usually a covalently closed piece of circular DNA (plasmid). Scientists have developed several methods to optimize the uptake of DNA by bacteria. This optimization process is called making the bacteria competent. The most common procedure for making bacteria competent involves incubation with $CaCl_2$ followed by mild heat. This makes the bacterial membrane leaky enough to allow foreign DNA to enter the cytoplasm. Another way is called electroporation, in which the bacterial cells are subjected to an electrical current, which also causes the cells to take up foreign DNA.

Transformations, among other applications, are used when cloning foreign DNA into a vector (ligation reaction). Figure 12.7 shows a schematic of a transformation from a ligation reaction. This ligation mixture is simply incubated with competent cells and then plated out on selectable media. The efficiency of a ligation reaction is low, so only a small percentage of the vectors will ligate to the foreign DNA, forming a circular piece of DNA. These new recombinant plasmids will be taken up by the competent cells. Bacteria cannot take up linear DNA, so most of the cells will not transform. These cells do not contain the Amp^r gene and thus will not grow in the presence of ampicillin. These are shown on the right. The cells shown on the left and center were transformed. Both took up a form of the plasmid and both will grow in the presence of ampicillin. The left side cells were transformed with the plasmid carrying the insert. The center cells were transformed with the plasmid that simply resealed without the insert. (*Note:* this can only happen if the vector has compatible ends.)

12.7   Selection

Selection is the process of determining which bacterial cells were correctly transformed with the vector/insert combination. It is the selectable markers that make this possible. Figure 12.7 shows the classical (i.e., old-fashioned) selection. If we assume that we have three types of cells in our mixture—the ones that have no plasmid, the ones with recircularized plasmid, and the ones with plasmid and insert—we can plate this mixture on a medium of tetracycline and a medium of ampicillin. The cells that didn't take up any plasmid will not grow on ampicillin or tetracycline because they do not carry the resistance that is conferred by the plasmid. The foreign DNA is cloned into the tetracycline gene, so a plasmid that has the insert will have an inactive tetracycline resistance gene but a functional ampicillin gene. These cells will grow on ampicillin but not on tetracycline. The cells that contain recircularized plasmid will grow on both media. The disadvantage to this selection process is that the cells you really want are the ones that are ampicillin resistant and tetracycline-sensitive. Thus you only found out which ones you wanted by killing them with tetracycline.

The plasmids incorporating the *lacZ* gene were designed to make selection simpler via blue/white screening as described in Section 12.2. Look at Figure 12.7 again. Imagine that instead of the tetracycline gene, we had the *lacZ* gene with the multiple cloning site within it. The foreign DNA would disrupt the *lacZ* gene. A recircularized plasmid would have a functional *lacZ* gene. If the cells that were transformed lacked the α-subunit of β-galactosidase, being transformed by recircularized plasmid would restore activity. If our transformation mixture were plated onto a medium containing ampicillin and the dye X-gal, we would quickly identify the colonies we want. They would be the white ones. The blue colonies would be the ones with the recircularized plasmid. Cells that took up no plasmid would not grow at all, of course.

12.8 Expression

The focus of this chapter is the **expression** or production of foreign proteins in bacteria. So far, we have described how to clone the foreign DNA into special expression vectors and the important consideration one needs to make when doing this. The overall purpose of this cloning is to produce the protein the DNA encodes. As described earlier, special cell lines and expression vectors are used. Figure 12.8 shows a common expression vector called pET 5a. As is the case with any plasmid, the pET 5a vector has an origin of replication and a **selectable** marker (Amp^r). It also has a MCS, which is on the right side of the circular map shown. In front of this multiple cloning site is the T7 polymerase promoter, which allows for transcription (the production of mRNA) of any gene that is placed in the MCS. Thus, in the presence of T7 RNA polymerase, the gene in the MCS will be transcribed and translated into protein.

This description of expression is an oversimplification, as there are many considerations for a successful expression. First, some proteins are toxic when expressed in *E. coli*. You don't want these proteins produced all the time, so it is desirable to turn on the expression of the gene (foreign DNA) only when necessary. The expression is regulated by controlling the production of T7 RNA polymerase in the host cell, which is itself under the control of a lactose inducible promoter (*lac* operon). If you control the amount of T7 polymerase, you can control the transcription from the T7 promoter. The *lac* operon is the operon that produces β-galactosidase when the cell needs to use the lactose as an energy source. A repressor protein is normally bound to the promoter preventing the host cell's RNA polymerase from transcribing the operon. Lactose is the natural inducer that binds the repressor removing it from the promoter and allowing transcription. Figure 12-9 gives an overview of the *lac*

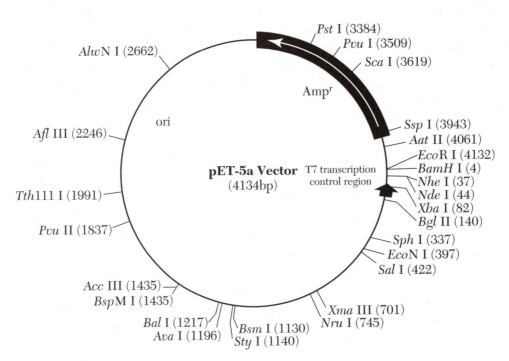

Figure 12.8 pET 5a expression vector

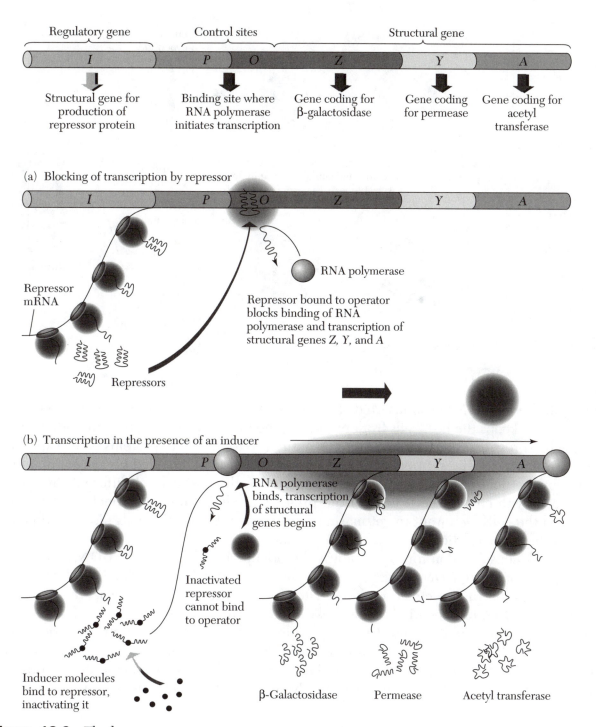

Regulatory gene | Control sites | Structural gene

| I | | P | O | Z | Y | A |

Structural gene for production of repressor protein

Binding site where RNA polymerase initiates transcription

Gene coding for β-galactosidase

Gene coding for permease

Gene coding for acetyl transferase

(a) Blocking of transcription by repressor

Repressor mRNA

RNA polymerase

Repressor bound to operator blocks binding of RNA polymerase and transcription of structural genes Z, Y, and A

Repressors

(b) Transcription in the presence of an inducer

RNA polymerase binds, transcription of structural genes begins

Inactivated repressor cannot bind to operator

Inducer molecules bind to repressor, inactivating it

β-Galactosidase   Permease   Acetyl transferase

Figure 12.9   The *lac* operon

operon. The molecule isopropyl-β-D-thiogalactopyranoside (IPTG) is a nonmetabolizable analog of lactose. When this molecule is added to culture medium, it binds and removes the *lac* repressor, and the system is induced.

Thus, when you induce the cells with IPTG, you induce the production of T7 RNA polymerase. This polymerase then finds the T7 promoter on the expression vector and transcribes

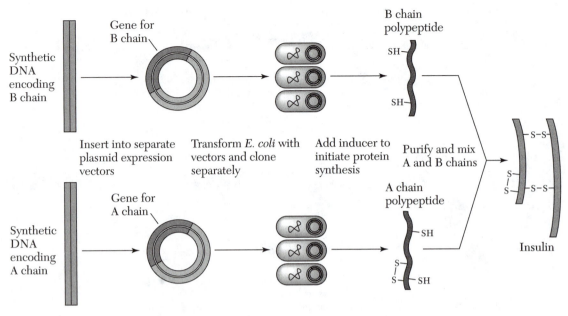

Figure 12.10 Cloning and expression of insulin

the gene (foreign DNA) you inserted. The host cell then translates the mRNA into protein. This system keeps the host cells from producing the foreign proteins until you induce them.

Each expression experiment needs to be optimized for different proteins being expressed, because every protein is different. The concentration of IPTG needed for maximum induction must be determined. The IPTG can be added to the host cells at different time points in their growth curve. The toxicity of the proteins produced must be checked. If the proteins kill the cells, you want to get maximum expression before the cells die. The time and temperature of induction is also a critical parameter.

Another consideration for expression is the orientation of the inserted DNA. If you cut out a piece of foreign DNA using *Nde* I at the 5′ end and *Xba* I at the 3′ end, you could insert it into the pET vector using those sites (see Figure 12.8). However, this would put the inserted DNA into the pET vector backward to the direction of transcription. If you used *Nde* I and *Eco*RI, then the insert would go into the pET vector in the proper orientation.

Remember that the ultimate goal is the isolation of large quantities of a protein that has interest to you. Figure 12.10 shows how cloning and expression have been used to produce mass quantities of insulin, which was formerly very expensive and difficult to produce.

12.9 Why Is This Important?

Millions of people suffer from diabetes, a disease in which too little insulin is produced. Insulin used to be produced only by purifying it from mammalian sources. This was expensive, time-consuming, and used many laboratory animals. With the invention of cloning technology, insulin can now be produced by using the replicative machinery of bacteria. It is cheaper and easier to get, improving the lives of those afflicted with the disease. As a pure

Expression of foreign proteins is the goal of a cloning experiment. If we want to obtain large quantities of a particular protein, we can clone the gene for that protein into a suitable expression vector. This vector can then be transformed into a host bacterial cell. The bacteria then produce the protein of interest for us, and we can purify it from there. Expression vectors and host cells are chosen so that we can turn on the production of the desired protein. This happens only at certain times in the growth of the bacteria. A common way is to have the transcription of the foreign gene controlled by T7 RNA polymerase, which is not normally present in the cell. The host cell is regulated so that it produces the T7 RNA polymerase under the control of the *lac* operon. In this way, when we give the lactose analog, IPTG, to the host cells, they produce the T7 polymerase. This then transcribes the foreign gene, and the host cell translates the mRNA into the protein we desire.

science tool, cloning is used to produce and study DNA and protein sequences. Scientists are also attempting to use cloning techniques to repair damaged genes that cause a wide variety of health problems. This is known as gene therapy and is a current and sometimes controversial topic.

Experiment 12

Cloning and Expression of Barracuda LDH-A

In this experiment, you will purify barracuda LDH DNA and clone it into a pET 5 expression vector. An expression cell line will then be transformed and the LDH expressed. Because this is a long experiment and some instructors may omit certain parts, the procedures and analysis will be divided into several parts.

Prelab Questions

1. What is meant by the term *cloning*?

2. What is meant by the term *expression*?

3. What are the two vectors we will use in this experiment?

4. What are the two *E. coli* cell lines that we will use?

5. Why are we using two different *E. coli* cell lines?

Objectives

The purpose of this experiment is to clone and express the gene for LDH from the barracuda, *Sphyraena lucasana*. In the process, the following ideas and techniques will be used:

Preparation of competent *E. coli* cells

Preparation of LB agar/ampicillin plates

Digestion of plasmid (vector) and foreign DNA with restriction enzymes

Agarose gel electrophoresis of DNA

Purification of DNA bands from agarose gels

Ligation of foreign and vector DNA

Transformation of competent cells and selection of transformants

Induction of protein expression

Verification of induction by SDS-PAGE and activity assays

Experimental Procedures: Part I

This section will include preparation of ampicillin plates, preparation of competent cells, digestion of plasmids with restriction enzymes, agarose gel separation of restriction fragments, and gel purification of plasmids and LDH DNA.

Tip 12.1

⊃ This experiment counts on many things you cannot control, such as whether the DNA ligase is still active, and many things you can control, such as whether you can successfully pipet 0.5 μL of a solution. This would be a good time to check your pipetting technique with quantities of 1 μL and smaller.

Day 1

We will be pouring ampicillin plates, which we will eventually use to screen for cloning and protein expression.

Preparation of Ampicillin Plates

1. Prepare 200 mL of L-Broth agar to pour plates with. Use 1-L flasks. L-Broth agar plates are the following: 0.5% bactoyeast extract, 1% NaCl, 1% tryptone, and 1.5% agar.

2. Make up the L-Broth agar in a 1-L flask, cover it with aluminum foil, place a piece of autoclave tape on the flask, and take it to the autoclave.

3. When the flasks are done autoclaving, place them into the 52°C water baths. Swirl the flasks occasionally while they are cooling. Swirl gently to keep air bubbles to a minimum.

4. When the flasks have cooled to the temperature of the water bath (15–30 min), add 100 μL of ampicillin (100 mg/mL) and swirl. If you can touch the flask for a few seconds without burning your hand, the medium is cool enough.

5. The plates can now be poured, with about 20 mL in each plate. Pour all the plates you can from your batch, even though you may not personally need them. Extras may come in handy later.

6. Let the plates sit at room temperature for 24 hr, then place them upside-down in a plastic bag and store in a dark refrigerator until needed. Ampicillin is light-sensitive.

Day 2

The goal for today is to cut the LDH gene (open reading frame) out of the PCRScript vector it is currently in and to cut open the pET vector that we will eventually put it into. We will be using two restriction enzymes, *Nde* I and *Eco*RI. While the enzyme digestions are going on, we will also prepare competent cells from *E. Coli* DH5α. Recall that these two enzymes leave sticky ends, but the two sites are not complementary.

Digestion of Plasmids and Barracuda LDH DNA

1. Set up a restriction enzyme digestion with *Nde* I and *Eco*RI for the PCRScript vector as shown. Remember to add the enzymes *last* and *don't remove them from the freezer block for more than 15 sec.*

PCRScript vector with LDH insert (1 μg)	10 μL
10× Buffer H	2 μL
ddH$_2$O	6.5 μL
Nde I (10 U/(L)	0.5 μL
*Eco*RI (10 U/(L)	0.5 μL
RNase (10 mg/mL)	0.5 μL
Final volume	20 μL

2. At the same time, set up a restriction enzyme digestion for the pET expression vector as shown:

pET expression vector (1 μg)	10 μL
10× Buffer H	2 μL
ddH$_2$O	6.5 μL
Nde I	0.5 μL
*Eco*RI	0.5 μL
RNase (10 mg/mL)	0.5 μL
Final volume	20 μL

   Note: The volumes used for the DNA samples depend on the concentration obtained when they were prepped. Therefore, you may have to change this protocol to reflect different initial concentrations.

3. Incubate at 37°C overnight. Freeze until needed.

Preparation of Competent Cells

In this part, you will prepare some *E. coli* cells that will be used for transformation later. The cell line is DH 5α (or JM 109), which is good for cloning. Eventually you will need competent JM 109 DE3 or BL21 DE3, which are expression cell lines. These can be prepared at the same time, prepared later, or purchased.

1. Acquire 100 mL of cells. These were prepared for you in the following manner: Glycerol stocks of cells were grown overnight in 50 mL of L-Broth medium. These were then used to inoculate 1 L in L-Broth and grown until the optical density at 600 nm was between 0.5 and 0.6. This takes about 3 h at 37°C from the time you add the 50 mL to the liter. They were then cooled to 4°C.

2. Spin the cells at 2000 × g for 15 min at 4°C. Use sterile 500-mL centrifuge bottles.

3. Gently resuspend the pellet in 1 mL of ice-cold, sterile 0.1 M $CaCl_2$. Gently add 24 mL of sterile 0.1 M $CaCl_2$ and resuspend while keeping the bottle cold in an ice bath. Spin again as in step 2.

4. Gently resuspend the pellet in 25 mL of ice-cold, sterile 0.1 M $CaCl_2$, keeping the bottle cold as before. Incubate on ice for 20 min. Spin again as above.

5. Gently resuspend the pellet in 4.3 mL of 0.1 M $CaCl_2$. Add 0.7 mL cold glycerol.

6. Aliquot 200 μL of cells into sterile cryo tubes and freeze quickly in a dry ice/EtOH bath. Store at $-70°C$.

Day 3

Gel Electrophoresis of Digestion Reactions

The purpose here is to isolate the DNA bands from the digestions so that we can later purify them. We should have bands for PCRScript vector (2961 bp), pET vector (4100 bp), and the LDH gene (1000 bp). However, as with any experiment, you might not obtain results exactly as expected. Have your TA or instructor help analyze the bands.

1. Make 40 mL of 1% w/v agarose in 1× TAE buffer (40 mM TRIS-acetate, 1 mM EDTA, pH 8.0). Heat in the microwave until dissolved (ca 30 s).

2. Remove and swirl the flasks. Heat for a few more seconds, if necessary, until the agarose is clear. Make sure the agarose solution doesn't boil over, because that will change the concentration of your gel.

3. Let the flask sit for about a minute until you can hold it without burning your hand.

4. Pour the agarose gel as before (Experiment 9a), insert the 8–10-tooth comb set as deep as possible with the spacing tool, and let harden. Assemble the apparatus and cover the gel with TAE buffer.

5. Add 5 μL of 5× loading buffer (50% glycerol, 0.3% bromophenol blue, 0.3% xylene cyanol in 5× TAE) to each of your digestion reactions. Mix well and spin in a microfuge.

6. Acquire 2 DNA ladder samples. One of these is made of *Hind*III fragments of λ DNA. The other is a commercial low-molecular weight ladder that starts at 2 kb. These may already have the tracking dye added or may not. If not, add the same tracking dye.

7. Heat the samples at 65°C for at least 2 min, spin, and place on ice.

8. Load the samples as you have before, splitting the digestion reactions between two lanes, and electrophorese at 100 V until the lead dye front has traveled three-fourths of the way down the gel or until the bands have migrated a sufficient distance. When the lead tracking dye is a little over halfway, put 10 μL of ethidium bromide (10 mg/mL) into each buffer chamber of the gel box. *Wear gloves at all times and be careful not to spill any of the solutions once EtBr has been added.*

9. To observe the bands, use a UV lightbox. *Wear gloves, goggles, and a face shield, because the UV light will burn your skin.*

10. Excise the LDH insert bands, which should have a size of about 1000 base pairs, and the pET vector bands, which should have a size of 4100. Avoid cutting out excess gel. Note that the λ DNA fragments should have the following sizes: 23.2, 9.4, 6.5, 4.3, 2.3, 2.0, and 0.56 kb. Low-molecular-weight ladders have sizes of 2 kb and smaller, depending on source.

Purification of Excised DNA with Qiagen Gel Extraction Kit

1. Tare a balance on a microfuge tube. Add the slices from the LDH DNA and reweigh the tube. Repeat for the pET vector. From now on, all procedures refer to both tubes.

2. Add three volumes of Buffer QG (i.e., 300 μL of buffer for 100 mg of gel).

3. Incubate at 50°C for 10 min. To help dissolve the gel, mix by flicking and inverting the tube two or three times.

4. If the QG buffer has turned pink, add 10 μL of 3 M sodium acetate, pH 5.0, to the samples and mix by inverting.

5. Place a QIAquick spin column into a 2-mL collection tube and load the sample.

6. Centrifuge for 1 min.

7. Drain the flow-through from the collection tube and return the spin column to the same collection tube. The flow-through should not have the DNA.

8. Add 0.5 mL of Buffer QG to the QIAquick column and centrifuge for one minute.

9. Add 0.75 mL of Buffer PE to the QIAquick spin column and centrifuge for one minute.

10. Drain the wash flow-through from the collection tube. Centrifuge the spin column for an additional 1 min to remove the last traces of Buffer PE.

11. Place the QIAquick spin column into a clean 1.5-mL microfuge tube.

12. To elute the DNA, add 50 μL of Buffer EB (10 mM TRIS-Cl, pH 8.5) to the middle of the column membrane and centrifuge for 1 min. Make sure the buffer covers the membrane before spinning.

13. Label the tubes well and store in a freezer.

Analysis of Results

Experiment 12: Cloning and Expression of LDH (Part I)

Data

Sketch or provide a picture of your agarose gel, showing the DNA ladder and phage λ DNA/*Hin*dIII fragments, your pET 5a lane, and your PCRScript/LDH lane.

Questions

1. Why is it necessary to make sure the LB-agar solution is cool enough before adding the ampicillin? What would your results look like if you added the ampicillin when the solution was too hot?

2. If you did add the ampicillin to the LB-agar when it was too hot as mentioned in question 1, when would be the first time in the experiment that you would know it?

3. Give two reasons why you should not pipet more of the restriction enzymes into the digestions than the amount given in the protocol.

4. Given that we are gel purifying the necessary DNA fragments anyway, why do we do the restriction digestions of PCRScript/LDH and pET 5a in separate tubes?

5. Why is the word *gently* emphasized so often in the procedure for making competent cells? What is the procedure doing to the cells?

6. How many bands did you expect to see in your digestion lanes on the agarose gel? How many did you see? Explain any discrepancies.

7. Before running samples on an agarose gel, we always heat the samples to 65°C for 2 min and then put the samples on ice. What is the point of this step?

Experimental Procedures: Part II

This part of the experiment will take you through the ligation of LDH DNA to pET 5a, and transformation of competent cells.

Day 4

Ligation of LDH DNA to pET Vector

In this part, you will ligate your barracuda LDH gene into the pET expression vector. You will also run a control to check for reannealing of the pET vector.

1. Set up reactions as follows:

	pET + LDH (μ)	pET w/o LDH (μL)
LDH insert from gel	6	0
10× ligation buffer	1.5	1.5
pET vector	3	3
T4 DNA ligase	1	1
ddH$_2$O	3.5	9.5

2. Ligate overnight at room temperature, or at 15°C if you have such an incubator.

3. Your ligations must be recovered the next day.

Note: Make sure your ligase buffer is fresh. It contains ATP that will go bad if the buffer goes through more than three or four freeze–thaw cycles.

Preparation of SDS-PAGE

We will eventually be using SDS-PAGE to help verify that we did induce a foreign protein. Today is a convenient time to make these gels for the future.

1. Make a 12% SDS separating gel and a 3–5% stacking gel as we did in Experiment 9c.

2. Store the gels (with the combs still in) at 4°C.

Day 5

Because ligation reactions go bad after one day, you need to come in the next day for the transformations.

Transformation of Competent Bacterial Cells (DH 5α or JM 109)

In this procedure, we will take the ligation mixture, which should have ligated LDH gene and pET vector, and get the *E. coli* DH 5α (or JM 109) cells to take it up. This is called transformation.

1. Check the hot block to make sure it is at 42°C.

2. Obtain the following:

    Sterile yellow pipet tips

Sterile microfuge tubes (4 total)

A tub of ice

3. Get your competent cells that you made previously and *immediately* put on ice. You are doing four transformations, each requiring 200 μL, so make sure you have enough (1 mL is fine), but not too many cells. If you thaw a tube and do not use it, you cannot use it again.

4. Aliquot 10 μL of DNA into sterile microfuge tubes as shown. You actually want four tubes, the fourth one being a "no DNA" control:

Tube 1: ligation that contains vector + insert

Tube 2: ligation that contains vector alone

Tube 3: a control plasmid, such as pBR322 or pUC18

Tube 4: no DNA at all

5. Set these four tubes aside and thaw the cells by warming in your hands. Once there are no more ice chunks, *immediately* aliquot 200 μL of cells into each of the four microfuge tubes.

6. Transfer the four tubes to the 42°C block and incubate for **exactly** 2 min.

7. Return the tubes to ice for 1 min.

8. Add 500 μL of LB media (without ampicillin) to each tube and incubate in the 37°C block for 40 min, inverting periodically.

9. After the incubation, spin the tubes in the microfuge for 4 min at 14,000 × g. Remove the supernatant by inverting the tube once quickly.

10. Resuspend the pellet in the remaining liquid using a P-200 sterile yellow tip. If there is no supernatant left, add 50 μL of LB. Pipet this onto a LB/Amp plate and spread on the plate. Ask if you don't know how to do this.

11. Let the plates soak for about 15 min. Place in a 37°C incubator upside down.

12. The plates need to be harvested after about 24 h or put into a refrigerator until it is time to harvest colonies.

Day 6

Preparation of Overnight Cultures

1. Sometime after 2 P.M. the day following transformation, recover your plates.

2. Choose three colonies from your pET + LDH plates and add a loopful to 10 mL of LB in sterile Falcon tubes. Add 10 μL of 100 mg/mL ampicillin.

3. Place in a 37°C. shaking incubator.

Analysis of Results

Experiment 12: Cloning and Expression of LDH (Part II)

Data

Describe the results of your four transformation plates:

pET 5/LDH ligation:

pET 5 w/o LDH:

control plasmid:

No DNA control:

Questions

1. Which plate had the most colonies on it? Is this what you expected? Why?

2. What does it mean if the pET 5 w/o LDH has colonies?

3. What does it mean if the no DNA control has colonies?

4. What are some possible products of the ligation reaction? Which ones are most likely?

5. After growing the transformed bacteria on the ampicillin plates, why do we select three colonies to prepare overnight cultures? Why do we add more ampicillin?

Experimental Procedures: Part III

In this part, you will verify that the colonies you grew up do contain a plasmid with inserted LDH DNA. The reason for doing this is to avoid wasting time trying to express something that might not be there. Often, despite our best efforts, plasmids reclose on themselves without the inserted gene.

Day 7

Verification of Clones: Plasmid Prep

There are two types of plasmid preps that can be run. One is a quick and cheap method, and the other is the more expensive Qiagen method.

Qiagen Plasmid Prep

1. Reclaim your three overnight cultures. Assuming they all grew, take 3 mL of each to do a plasmid prep. Store the rest at 4°C.

2. Split the 3 mL into two microfuge tubes and spin for 2 min.

3. Resuspend the bacterial pellets in 0.3 mL of Buffer P1, which should have had RNase A added already.

4. Add 0.3 mL of Buffer P2, mix gently by inversion four to six times, and incubate at room temperature for 5 min.

5. Add 0.3 mL of chilled Buffer P3, mix immediately and gently by inversion four to six times, and incubate on ice for 5 min.

6. Mix the sample again and centrifuge it at maximum speed in a microfuge for 10 min. Remove and save the supernatant promptly. The supernatant should be clear.

7. Set up a Qiagen tip 20 in a rack and equilibrate by applying 1 mL of Buffer QBT. Allow the column to empty by gravity flow.

8. Apply the supernatant from step 6 to the Qiagen tip 20 and allow to drain by gravity flow.

9. Wash the Qiagen tip 20 with 4 × 1 mL of Buffer QC.

10. Elute the DNA into a microfuge tube using 0.8 mL of Buffer QF. Drain by gravity flow.

11. Precipitate the DNA with 0.56 mL of room temperature isopropanol. Centrifuge at maximum speed for 30 min and carefully decant the supernatant.

12. Wash the DNA with 1 mL of 70% ethanol and centrifuge for 10 min.

13. Decant the supernatant and dry the pellet. Redissolve in 10 μL of TE buffer.

Quick and Cheap Plasmid Prep

1. Using a P-1000 tip, transfer 1.5 mL of O/N culture to a 1.5 mL microfuge tube. Spin down at 14,000 × g for 3 min. Decant the supernatant, making sure the cells have pelleted on the bottom of the tube.

2. Resuspend the cells in 100 μL of TE buffer (10 mM TRIS, 1 mM EDTA, pH 8.0) using a P-200 tip. Make sure the pellet is completely resuspended.

3. Add 200 μL of 0.2 M NaOH, 1% SDS. Vortex the tube on high for 5 s and put it on ice for exactly 5 min. The 5-min incubation is critical. If you incubate too long, you will denature your DNA.

4. Add 150 μL of cold 5 M KOAc. Vortex the tube upside down. Put it back on ice for 5–10 min.

5. Centrifuge the tube for 5 min at 14,000 $\times$ g. Transfer the supernatant to a clean microfuge tube. The pellet should be white and smeared on the side of the tube. Try not to transfer any of this to the clean tube. The longer you let the pellet sit before transfer, the looser the pellet becomes.

6. Add 1 mL of ice-cold EtOH to the transferred supernatant. Incubate this in a freezer for 10 min. Invert the tube several times. Centrifuge for 15 min at 14,000 $\times$ g. Sometimes you cannot see a pellet after this spin, so put the tubes in the centrifuge facing in the same direction so that you will know where the pellets are.

7. At the end of the spin, you should have a small, white pellet. This is your DNA. Decant the EtOH, making sure not to lose the pellet. To get rid of the excess EtOH, place the open tube in a 37°C hot block for 10 minutes.

8. Resuspend the tube in 15 μL of H_2O. Add the water to the side of the tube and close. Flick the water down and let the tube sit for 5 min.

Note: This prep will have lots of RNA and protein in it. It can be used for transformations and digestions. For digestions, RNase A should be added to the reaction.

Restriction Digestion

Digest 2 μL and 5 μL of your plasmid prep with *Eco*RI and *Nde* I overnight. Set this up for a total of 20 μL per digestion. Be sure to save the rest of the plasmid prep.

Day 8

Electrophoresis of Digested Plasmid

1. Acquire the λ DNA fragment ladder and low-molecular-weight ladder you will use as markers.

2. Prepare a 1% agarose gel.

3. Add 5$\times$ tracking dye to the digestions and the DNA ladders, heat at 65°C for at least 2 min, and then place on ice.

4. Load the samples and electrophorese at 80–100 V.

5. When the lead tracking dye has moved about halfway, add 5 μL of ethidium bromide (10 mg/mL) to each buffer chamber.

6. You can examine your gel at any time to see if the bands have separated well enough to tell you if you had a plasmid with insert.

Analysis of Results

Experiment 12: Cloning and Expression of LDH (Part III)

Data

1. Sketch or provide a picture of your agarose gel, showing the digested pET 5/LDH and the molecular weight markers.

2. Describe the number and location of the bands.

3. Which plasmid prep did you do and why?

Questions

1. What is the purpose of the verification of clones? Why do we not just scale up a colony that grew after the transformation and transform that into the expression cell line?

2. What would it indicate if you ran your restriction-digested plasmid prep on an agarose gel and saw three bands (one at 5000, one higher, and one lower) in the lane that was supposed to have pET 5/LDH?

3. What would it indicate if you ran your restriction-digested plasmid prep on an agarose gel and saw one band at 5000 in the lane that was supposed to have pET 5/LDH?

4. How could you handle the situation described in question 3?

Experimental Procedures: Part IV

In this part we will finish the expression of barracuda LDH.

Day 9

Expression of LDH

The growth and induction portion involves the following: Your successful plasmid prep is used to transform BL 21 DE3 (or JM 109 DE3) cells, and an overnight culture is grown. Your overnight cultures are taken and put in larger volumes of LB-Broth with ampicillin. This is known as a scale-up. When the cells are finally being grown in an appropriate volume for the amount of protein you eventually want, they are grown to an optical density of 0.4. Some of the cells are then taken as controls: the others are induced with IPTG at a final concentration of 1 mM. Both are allowed to grow for 2 h at 30°C. You may be asked to do some of this, or it may be done for you.

Determination of LDH Activity

If you are going to check the induced cells for LDH activity compared to the controls, follow these procedures:

1. When the induction is over, the cells are spun down in preweighed bottles or vials at 5000 rpm at 4°C. The supernatant is discarded and the cell weight is measured. The cells are now ready to be used or frozen at $-70°C$.

2. Resuspend the cells in 10 volumes of bicine buffer (0.03 M bicine, pH 8.5). Keep everything on ice.

3. Sonicate the cell suspension for 30 s and then allow the cells to cool on ice for 30 s. Repeat this sonication cycle three times.

4. Centrifuge the cell lysate at 10,000 rpm for 15 min at 4°C. Remove the supernatant.

5. The supernatants are now ready for LDH activity assays as in Experiments 4–8.

SDS-PAGE

If all you are going to do is check for induction via SDS-PAGE, here is an easier way to proceed.

1. After induction, take 1 mL of cells and spin down at $14,000 \times g$ for 5 min.

2. Resuspend the pellet in 50 μL of 1× SDS loading buffer (Chapter 9).

3. Load and electrophorese 10 μL on a 12% SDS-PAGE as in Chapter 9.

4. Stain the gel in Coomassie Blue and destain in 10% acetic acid as in Chapter 9.

5. *Congratulations!* If everything went according to plan, you have just successfully cloned a barracuda gene for LDH into a bacterial host and had the bacteria express it.

6. Of course, you may not be quite done yet. Chapter 13 involves the PCR reaction to amplify the barracuda DNA.

Additional Problem Set

1. Explain and diagram how blue/white screening works. How does this compare to the earlier selections using ampicillin and tetracycline resistance markers?

2. What are the general requirements for a plasmid used in cloning?

3. What are the general requirements for a plasmid used in expression of foreign proteins?

4. Outline the procedures you would use to produce human erythropoietin in bacteria. What might such a compound be used for?

5. The gene for β-globin is a split gene, which means it contains introns. How would that affect a plan to clone human β-globin in bacteria?

6. If you wanted to know the sequence of a protein, what would be the advantages and disadvantages of sequencing the protein directly vs. sequencing the DNA for the protein?

7. A vector has a polylinker (MCS) containing restriction sites in the following order: *Hind*III, *Sac* I, *Xho* I, *Bgl* II, *Xba* I, and *Cla* I.
   a. Give a possible nucleotide sequence for the polylinker.
   b. The vector is digested with *Hind*III and *Cla* I. A DNA segment contains a *Hind* III restriction site 650 bases upstream from a *Cla* I site. This DNA segment is digested with *Hind*III and *Cla* I, and the resulting *Hind*III–*Cla* I fragment is directionally cloned into the digested vector. Give the nucleotide sequence at each end of the vector and the insert, and show that the insert can be cloned into the vector in only one orientation.

Webconnections

For a list of Web sites related to the material covered in this chapter, go to **Webconnections** at the *Experiments in Biochemistry* site on the Saunders College Publishing Web page. You can access this page at http://www.saunderscollege.com. **Webconnections** are under *Experiments in Biochemistry* in the Biochemistry portion of the Chemistry page.

References and Further Reading

K. W. Adolph, *Advanced Techniques in Chromosome Research,* Marcel Dekker, 1991.

F. M Ausubel, R. Brent, R. Kingston, et al. *Current Protocols in Molecular Biology,* Wiley, 1987.

R. F. Boyer, *Modern Experimental Biochemistry,* Addison-Wesley, 1993.

J. Beugelsdijk, *Automation Technologies for Genome Characterization,* Wiley Interscience, 1997.

M. K. Campbell, *Biochemistry,* Saunders College Publishing, 1998.

D. L. Crawford, H. R. Constantino, and D. A. Powers, (1989), *Mol. Biol. Evol.* **6(4).** Lactate Dehydrogenase-B cDNA from the Teleost *Fundulus heteroclitus:* Evolutionary Implications.

R. H. Garrett and C. M. Grisham, *Biochemistry,* Saunders College Publishing, 1999.

L. Z. Holland, M. McFall-Ngai, and G. N. Somero, (1997), *Biochemistry,* **36.** Evolution of Lactate Dehydrogenase-A Homologs of Barracuda Fishes from Different Thermal Environments.

D. Huang, C. J. Hubbard, and R. A. Jungmann, (1995), *Mol. Endocrinol.* **9(8).** Lactate Dehydrogenase-A Subunit Messenger RNA Stability is Synergisticaly Regulated via the Protein Kinase A and C Signal Transduction Pathways.

M. Maekawa, K. Sudo, S. S. Li, and T. Kanno, (1991), *Hum. Genet.* **88(1).** Genotypic Analysis of Families with Lactate Dehydrogenase A (M) Deficiency by Selective DNA Amplification.

H. Miyajima, T. Shimizu, and E. Kaneko, (1992), *Rinsho Shinkeigaku,* **32(10).** Gene Expression in Lactate Dehydrogenase-A Subunit Deficiency.

Promega Corporation, *Technical Manual for pET-5 Expression Vectors,* 1995.

J. M. Quattro, H. A. Woods, and D. A. Powers, (1993), *Proc. Natl. Acad. Sci.* **90(1).** Sequence Analysis of Teleost Retina–Specific Lactate Dehydrogenase C: Evolutionary Implications for the Vertebrate Lactate Dehydrogenase Gene Family.

F. S. Sakai, Sharief, Y. C. Pan, and S. S. Li, (1987), *Biochem. J.* **248(3).** The cDNA and Protein Sequences of Human Lactate Dehydrogenase B.

H. A. Wyckoff, J. Chow, T. R. Whitehead, and M. A. Cotta, (1997), *Current Microbiology,* **34.** Cloning, Sequence, and Expression of the L-(+) Lactate Dehydrogenase of *Streptococcus bovis.*

W. Zhou and E. Goldberg, (1996), *Biol. Reprod.* **54(1).** A Dual-Function Palindromic Sequence Regulates Testis-Specific Transcription of the Mouse Lactate Dehydrogenase c Gene In Vitro.

Chapter 13

Polymerase Chain Reaction

Topics

Introduction

This final chapter deals with the exciting and popular technique of polymerase chain reaction, a method for amplifying DNA in microgram amounts from picograms or less of starting material.

13.1   Amplification of DNA

In the previous chapters, we discussed how cloning could be used to increase the amount of a specific gene. With cloning, we let the host cell's DNA replication system amplify our target DNA for us. This chapter discusses another method used to amplify DNA without having to go through the cloning process, **polymerase chain reaction (PCR).**

PCR technology makes it possible to amplify a gene without cloning it from the rest of the DNA it was with. This target gene could be in a plasmid, in a virus, or in a chromosome. PCR is so sensitive that it can produce over 1 μg of specific target DNA from less than 50 ng in a few hours.

PCR is an automated procedure carried out in a machine called a **thermocycler** that controls the time and temperature of the amplification reactions. The key to the process was the discovery of a heat-stable DNA polymerase from the bacteria *Thermus aquaticus*.

This enzyme allows the reactions to proceed at high temperatures without denaturing the key enzyme that replicates the DNA. The other components of the reaction include the target DNA you want to amplify, the full set of deoxynucleoside triphosphates (dNTPs), and specific primers that are complementary to the DNA in or near the target DNA.

13.2 *Taq* Polymerase

Many enzymes are involved when an organism replicates its DNA, and the reaction occurs at biological temperatures (i.e., 20–40°C.). Some of these enzymes are involved in the synthesis and subsequent proofreading and repair of the new DNA: some are involved in the unwinding of the DNA double helix so that the replication can proceed. The complexity of this process and the requirement for ambient temperatures so that the enzymes could work made it impossible to automate a DNA replication outside of the organism. If we wanted to produce DNA chemically, we would have to find a way to separate the DNA strands first. This would require a temperature that would inactivate the average DNA polymerase that would be used to replicate the single strand.

In 1993, Dr. Kary Mullis won the Nobel Prize in science for his PCR procedure. He used the DNA polymerase from *T. aquaticus,* known now as Taq polymerase. The bacteria were found living in undersea vents at temperatures around 100°C. Any enzymes from these bacteria would have to work at high temperatures. Using Taq polymerase allowed the automation of PCR. A thermocycler can raise the temperature high enough to separate the strands of DNA without killing the enzyme. Then the thermocycler reduces the temperature, allowing the primers to anneal. The temperature is then raised again to the one optimal for the polymerase to work. The polymerase then copies the single strands starting at the primer and going from 5′ to 3′ as all DNA synthesis does. Once enough time has elapsed to replicate the required DNA sequence, the thermocycler raises the temperature again to separate the DNA strands. Each round of replication doubles the amount of DNA. It takes about 1 min to replicate a kilobase of DNA. This means that with this process, if your target DNA sequence is 1 kb, you can double it every couple of minutes. At that rate, you can get a huge amplification in a few hours. Figure 13.1 demonstrates the basics of this reaction.

The shaded DNA in the middle is the target sequence. The reaction begins with the separation of the strands. A common way to do this would be to heat the samples to 95°C for 30 s. In step 2, the reactions are cooled to 60°C for 30 s. This allows the primers to anneal. The primers are shown as the short DNA sequence that is attached just outside of the target sequence. These primers are always present in excess so that as soon as DNA is separated, the primers are there, ready to bind. In the third step, the samples are heated to 72°C to allow the Taq polymerase to extend the DNA starting with the primers and heading towards the 3′ end. That is the end of one cycle. This cycle is repeated as often as necessary to obtain the desired amount of target DNA. It is important to note that the amount of primer and the amount of dNTP actually control the maximum possible PCR product. Once you run out of either, the reaction stops.

One of the problems associated with PCR is that by using the artificial system of replication, we lose the proofreading and repair functions that many organisms have as part of their DNA replication. Taq polymerase makes a high number of errors when it synthesizes DNA. If one of these errors happens early, that mistake is amplified many times, and the product you get will be different from the target DNA you started with. DNA sequencing

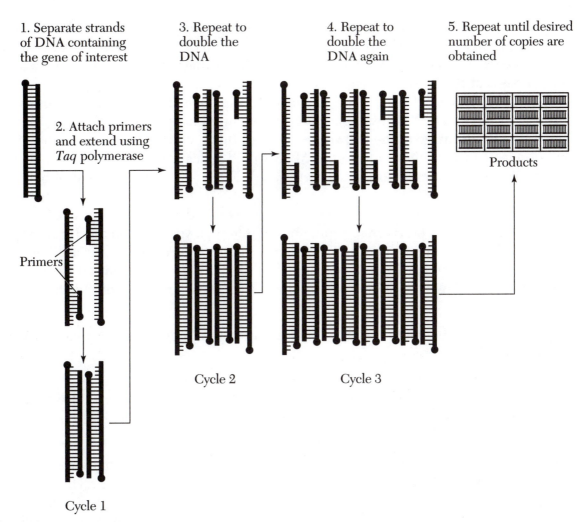

1. Separate strands of DNA containing the gene of interest

2. Attach primers and extend using *Taq* polymerase

3. Repeat to double the DNA

4. Repeat to double the DNA again

5. Repeat until desired number of copies are obtained

Primers

Products

Cycle 1

Cycle 2

Cycle 3

Figure 13.1   The polymerase chain reaction (PCR)

is often necessary to verify that PCR product is correct. Other heat-stable DNA polymerases have since been discovered, and some additional enzymes that confer a proofreading capability, such as **vent polymerase,** are also used.

13.3   Primers

Designing primers is the most important part of a successful PCR. Figure 13.1 shows the primers binding outside the target DNA. This is one way to do it. Another way is to have the primers be part of the target DNA sequence. In either case, the target DNA is replicated.

Primers are 18–40 bases long. If a primer is too short, it will not bind specifically enough to the region of DNA you want. If it is too long, it becomes expensive and the annealing time becomes longer. Each strand of DNA must have its own primer, called the forward and reverse primers. These will have different sequences because they are designed to bind to different pieces of DNA. It is important to make sure the primers won't anneal to one another.

Custom primers are those that you specify and order for your particular PCR needs. For example, if we know we are trying to do PCR on the barracuda LDH-A gene, and we know its sequence, we can specify the exact bases we want that would be complementary to the first 25 bases at the 5′ end of each strand.

Standard primers are those that are used with common vectors. If you have your gene of interest in pBlueScript™, for example, and the gene was cloned into the multiple cloning site, you can take advantage of the known sequence of this vector on each side of the MCS. Primers designed to the DNA outside the MCS would bracket your target DNA, so that anything between these primers would be replicated. Figure 13.2 shows a map of how this might work.

Outside of the MCS, you can find known sequences based on the T3 RNA polymerase promoter and the T7 RNA polymerase promoter. Therefore, if you ordered the T7/T3 primer set, you would be able to replicate via PCR whatever target DNA you had inserted into the multiple cloning site. Another common primer is the M13 primer, also shown in Figure 13.2.

Primers must be chosen or designed with great care. These are some of the characteristics of good primers:

1. The primers must be long enough to be specific to the target DNA or that DNA just surrounding the target.

2. The primer must be short enough to meet your budgetary concerns. Clearly, if you could make a 1000-base primer, it would be very specific, but you probably would have synthesized the entire gene. If you could afford to do that, and it were possible, you wouldn't need to use PCR.

3. The forward and reverse primers should have similar G and C contents. Remember that the melting temperature of DNA is dependent on the G + C content. Thermocycler programs are optimized for annealing temperatures and times. If the forward and reverse primers have radically different G + C contents, then the same program will not work well for both primers.

4. The primers should have low complementarity to each other. You want the primers to bind to the target DNA or that which surrounds it. If your forward primer is 5′-AAATT-TAAATTT-3′ and your reverse primer is 5′-TTTAAATTTAAA-3′, then your two primers would just as soon bind to each other as to the target DNA. This would remove your primers from the reaction and you would get no product.

5. Primers should have minimal secondary structure. The primers shown in observation 4 are bad for another reason. Each one of them would easily bind to itself, making a hairpin loop:

5′-AAATTT ⌐
3′-TTTAAA ⌐

This would also eliminate the primer from binding with the target, so your reactions would not continue.

◄► Practice Session 13.1

If you had the LDH gene cloned in pBluescript as shown here, what primers could you use to amplify the gene by PCR? The LDH gene was inserted into the vector with *Bam*HI and

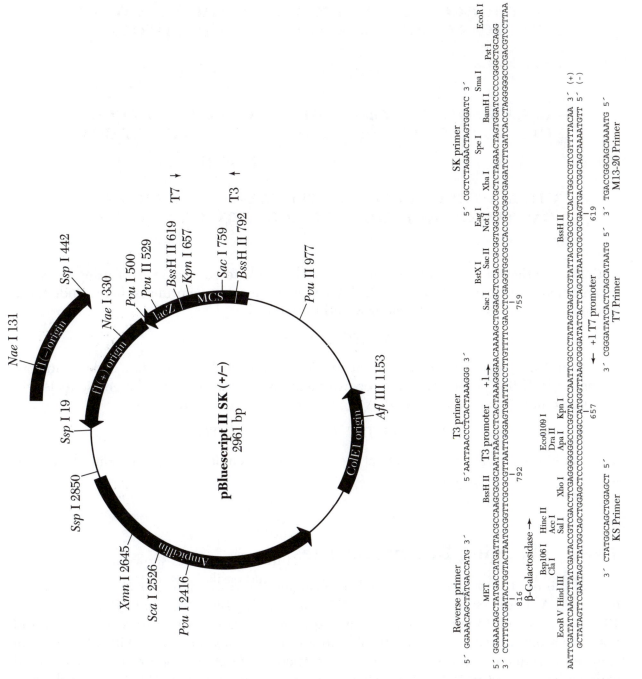

Figure 13.2 pBluescript and partial sequence

*Hind*III. The vector sequence is shown in italics. The target DNA sequence is shown in normal type.

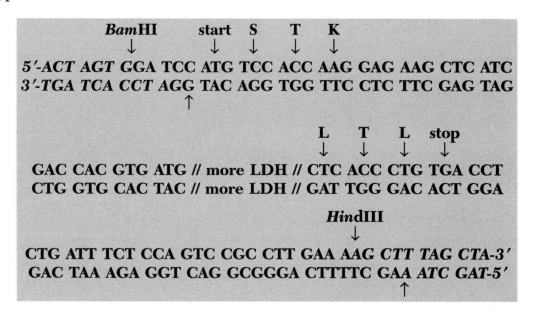

```
        BamHI          start    S     T     K
          ↓              ↓       ↓     ↓     ↓
5'-ACT AGT GGA TCC ATG TCC ACC AAG GAG AAG CTC ATC
3'-TGA TCA CCT AGG TAC AGG TGG TTC CTC TTC GAG TAG
                    ↑


                            L     T     L    stop
                            ↓     ↓     ↓     ↓
GAC CAC GTG ATG // more LDH // CTC ACC CTG TGA CCT
CTG GTG CAC TAC // more LDH // GAT TGG GAC ACT GGA

                          HindIII
                             ↓
CTG ATT TCT CCA GTC CGC CTT GAA AAG CTT TAG CTA-3'
GAC TAA AGA GGT CAG GCGGGA CTTTTC GAA ATC GAT-5'
                                          ↑
```

Let's assume that we want to have our PCR product able to go into a vector using the same two restriction sites, *Bam*HI and *Hind*III. We would want to have primers that include these sequences. A possible forward primer would therefore be

5'-GGA TCC ATG TCC ACC AAG GAG-3'

This would match the restriction site at the 5' end and include the start codon and four amino acids. If this primer proved to be insufficiently specific, you could increase its length by adding more bases at the 3' end. A possible reverse primer would be

5'-AAG CTT TTC AGG GCG GAC TGG-3'

This would retain the restriction site at the 3' end and include several noncoding bases that were part of the LDH gene.

13.4 Changing Restriction Sites

One of the most powerful uses for PCR is changing the restriction sites at the ends of the target DNA. Let's assume we cut the LDH gene out of pBluescript as in Practice Session 13.1. If we wanted to put the gene into an expression vector, but the expression vector did not have a *Bam*H1 site, would we be able to use it? The answer is yes. With PCR, we can change the sequence and put in a different restriction site. A common restriction site used with expression vectors is *Nde* I, which recognizes the sequence CAT ATG. One of the reasons this is so common with expression vectors is that it makes use of the start codon. Every gene has to have an ATG sequence, and by using *Nde* 1 you eliminate upstream noncoding regions.

If we wanted to clone our LDH gene into an expression vector using an *Nde* I site, all we would have to do is change our primers to include the *Nde* I sequence. Then PCR would produce a product with the proper ends. Instead of the primers we used before, we would use the following:

CAT ATG TCC AC AAG GAG

In the first cycle, the CAT would have nothing to bind to because the 3′ strand is not complementary at that position:

CAT -ATG TCC ACC AAG GAG
TGA TCA CCT AGG TAC AGG TGG TTC CTC

However, there is enough of the primer that is complementary to make it bind, and the CAT would just go along for the ride on the first cycle. After the first cycle, the replication of the 5′ strand would create the complement to the CAT, such that the 3′ strand would start as follows:

3′-GTC TAC AGG TGG TTC CTC

Now you have a PCR product that begins with the *Nde* I site. This method is also used to change individual bases within the gene product, in a powerful technique called **site-directed** mutagenesis.

ESSENTIAL INFORMATION

Polymerase chain reaction (PCR) is a very powerful technique that allows the rapid and specific production of target DNA. It is not based on cloning technology, and no organisms need be grown to produce the DNA. The target DNA can be free or in a suitable vector. PCR is possible due to the discovery of heat-stable forms of DNA polymerase. This allows an automated reaction sequence of dissociation of double-stranded DNA, binding of primers, and extension with the polymerase. The design of the primers is the most critical part of PCR. The primer must bind at the right place on or near the target DNA, must not bind to itself, and must have the proper length to meet specificity and cost considerations. Using PCR technology, individual bases can be changed from the orginal target DNA. This might be done to change the restriction sites at the end of a gene you are trying to insert into another vector, or it might be done to mutate the actual DNA sequence of the gene.

13.5 Why Is This Important?

PCR technology is quickly becoming the most important single biotechnology technique that you could learn. It is amazingly simple once you have acquired the correct primers and learn how to program the thermocycler. With PCR, almost nonexistent samples of DNA can be amplified to give microgram quantities for analysis. In many labs, PCR is replacing classical cloning procedures for scale-ups of DNA. With PCR, individual bases can be changed in the important technique of site-directed mutagenesis. These bases might be changed to allow the DNA to be inserted more easily into a vector of choice. They might be changed to study the resulting structure/function differences of the protein translated. Forensics uses PCR to obtain enough DNA from a crime scene to eventually identify or exclude a potential criminal.

Experiment 13

PCR of Barracuda LDH-A

In this experiment, you will use another technique to amplify the DNA from barracuda LDH, namely polymerase chain reaction (PCR).

Prelab Questions

1. How is PCR different from cloning?

2. What is meant by a "forward" or a "reverse" primer?

3. What are characteristics of a good primer?

4. What are our primers complementary to?

5. What is the purpose for each of the steps in the thermocycler routine?

Objectives

The purpose of this experiment is to use PCR to amplify the gene for LDH from the barracuda, *Sphyraena lucasana*.

Experimental Procedures

Standard Procedure

1. Set up three reactions as shown. These will include a negative control with no DNA template added, a positive control using pUC 18 to verify that the PCR reaction worked with known primers, and a reaction with the PCRScript/LDH plasmid. It is best to use a cocktail for much of this.

	No DNA Control	pUC 18 DNA	PCRScript
DNA template	0	1 μL	1 μL
Primer 1	5 μL	0	5 μL
Primer 2	5 μL	0	5 μL
Control primer 1	0	5 μL	0
Control primer 2	0	5 μL	0
Taq polymerase	0.5 μL	0.5 μL	0.5 μL
dNTPs	16 μL	16 μL	16 μL
10× PCR buffer	10 μL	10 μL	10 μL
ddH$_2$O	63.5 μL	62.5 μL	62.5 μL

The forward primer is the following:

5'-C CAT ATG TCC ACC AAG GAG AAG CTC ATC GAC CAC GTG ATG

The reverse primer is the following:

5'-TTC AGG GCG GAC TGG AGA AAT CAG AGG

The control primers are standard M13 forward and reverse.

Note: Volumes are adjusted to give 100 pmol of each primer and 1–100 pmol of template, depending on the source. We used 1 pmol. The dNTP stock is 2 mM, and the Taq or Vent stock is 5 U/μL.

2. The program we will use for the thermocycler will do the following cycles:
   a. start at 95°C for 5 min (denaturation)
   b. 30 cycles of the following:
      i.   95°C for 1 min (denaturation)
      ii. 55°C for 1 min (annealing primers)
      iii. 72°C for 1 min (extension of primers)
   c. 4°C overnight

PCR Bead Method

Another way to set up PCR is to use PCR beads, which have all of the enzymes, buffers, and dNTPs necessary for the reaction. You just have to add your primer and template.

1. Set up the protocol shown below for reactions using PCR beads:

	No DNA Control	pUC 18 DNA	PCRScript
DNA template	0	1 μL	1 μL
LDH primer 1	5 μL	0	5 μL
LDH primer 2	5 μL	0	5 μL
M 13 primer 1	0	5 μL	0
M 13 primer 2	0	5 μL	0
ddH$_2$O	15 μL	14 μL	14 μL

2. If the thermocycler does not have a hot lid, add 50 μL of mineral oil to the tubes and start the same program as shown in step 1.

Gel Electrophoresis of PCR Products

1. Prepare a 1.25% agarose gel in TAE buffer. This is similar to the gels we made in Experiment 12, but the agarose concentration is a little higher.

2. Remove 10 μL of the PCR reactions (from underneath the mineral oil, if necessary) and add 2 μL of 6× loading buffer. Heat at 65°C for 5 minutes and then place on ice.

3. Run lanes with λ DNA/*Hin*d III fragments and low-molecular-weight DNA markers as in Experiment 12 (also heated to 65°C first).

4. Load the gel and electrophorese at 80 V.

5. Soak the gel in 50 mL of buffer or water with 10 μL of 10 mg/mL ethidium bromide for 15 min. Photograph on a UV lamp.

Analysis of Results

Experiment 13: Polymerase Chain Reaction of LDH

Data

1. Sketch or provide a picture of your agarose gel, showing the MW markers, control reaction, and LDH bands.

2. What are the sizes of the fragments generated by PCR?

Questions

1. What is the purpose of using the pUC18 plasmid and M13 primers?

2. What is the purpose of having the no DNA lane?

3. Why is it not possible to have automated PCR without a heat-stable form of DNA polymerase?

4. If you had no heat-stable form of DNA polymerase but wanted to do PCR, what would you have to do each cycle?

5. Using the primers listed in the procedures, could you take the PCR product and ligate it into a pET vector? Why or why not?

Additional Problem Set

1. Suppose that you are a prosecuting attorney. How has the introduction of PCR changed your job?

2. Why is DNA evidence more useful as exclusionary evidence than for positive identification of a suspect?

3. What difficulties arise in the polymerase chain reaction if there is contamination of the DNA that is to be copied? What would determine if the presence or absence of other DNA sequences in the reaction would affect your results?

4. Why are primers usually between 30 and 40 bases long?

5. Why should the G + C content of the forward and reverse primers be similar?

6. If you wanted the LDH reverse primer from Experiment 13 to have an *Eco*RI site at the end (farthest from the LDH gene), what site-directed mutagenesis could you do?

7. What controls the total number of copies of LDH that is amplified with PCR?

Webconnections

For a list of Web sites related to the material covered in this chapter, go to **Webconnections** at the *Experiments in Biochemistry* site on the Saunders College Publishing Web page. You can access this page at http://www.saunderscollege.com. **Webconnections** are under *Experiments in Biochemistry* in the Biochemistry portion of the Chemistry page.

References and Further Reading

M. Campbell, *Biochemistry,* Saunders College Publishing, 1998.

P. Cohen, *New Scientist, Ghosts in the Machines,* 1998.

R. H. Garrett and C. M. Grisham, *Biochemistry,* Saunders College Publishing, 1999.

L. Z. Holland, M. McFall-Ngai, and G. N. Somero, (1997), *Biochemistry,* **36.** Evolution of Lactate Dehydrogenase–A Homologs of Baracuda Fishes from Different Thermal Environments.

H. H. Jung, R. L. Lieber, and A. F. Ryan, (1998), *Am. J. Phys.* **275(1).** Quantification of Myosin Heavy Chain mRNA in Somatic and Branchial Arch Muscles Using Competitive PCR.

R. P. Oda, M. A. Strausbauch, A. F. R. Huhmer, N. Borson, S. R. Jurrens, J. Craighead, P. J. Wettstein, B. Eckloff, B. Kline, and J. P. Landers, (1998), *Anal. Chem.* **70.** Infrared-Mediated Thermocycling for Ultrafast Polymerase Chain Reaction Amplification of DNA.

Promega Corporation, *Technical Manual for pET-5 Expression Vectors,* 1995.

M. Verhaegen and T. K. Christopoulos, (1998), *Anal. Chem.* **70.** Quantitative Polymerase Chain Reaction Based on a Dual-Analyte Chemiluminescence Hybridization Assay for Target DNA and Internal Standard.

K. O. Voss, K. P. Roos, R. L. Nonay, and N. J. Dovichi, (1998), *Anal. Chem.* **70.** Combating PCR Bias in Bisulfite-Based Cytosine Methylation Analysis. Betaine-Modified Cytosine Deamination PCR.